1978

THE HAMILTON-JACOBI EQUATION

A Global Approach

This is Volume 131 in
MATHEMATICS IN SCIENCE AND ENGINEERING
A Series of Monographs and Textbooks
Edited by RICHARD BELLMAN, *University of Southern California*

The complete listing of books in this series is available from the Publisher
upon request.

THE HAMILTON-JACOBI EQUATION
A Global Approach

STANLEY H. BENTON, Jr.

Department of Mathematics
Montclair State College
Upper Montclair, New Jersey

ACADEMIC PRESS   New York   San Francisco   London   1977

A Subsidiary of Harcourt Brace Jovanovich, Publishers

ACADEMIC PRESS, INC.
111 Fifth Avenue, New York, New York 10003

United Kingdom Edition published by
ACADEMIC PRESS, INC. (LONDON) LTD.
24/28 Oval Road, London NW1

Library of Congress Cataloging in Publication Data

Benton, Stanley H
      The Hamilton–Jacobi equation.

      (Mathematics in science & engineering ; vol. 131)
      Bibliography: p.
      1.   Hamilton–Jacobi equations.   I.   Title.   II.   Se-
ries.
QA374.B34   515$'$.35   76-45986
ISBN 0–12–089350–9

To Jackie for everything

Contents

Preface

The original purpose of this work was to present the existence theorems of the author in a more accessible framework than the original research papers. Chapter II has been devoted to this purpose. However, the uniqueness question is of equal importance and is very subtle and difficult. Chapter III is devoted to this area, including the major uniqueness theorems of Douglis and Feltus. The whole area of global solutions is rather difficult to motivate without a background in the classical treatments of partial differential equations (PDEs); chapter I presents a brief introduction to the classical methods and may be skipped by the reader already familiar with this field. Chapter IV is included to offer a brief guide to the literature of the applications of the Hamilton–Jacobi equation and of numerical methods for solving this equation.

The whole area of global solutions of PDEs is undergoing rapid growth and enjoying wide interest, and the field has grown too vast to be included in one book. The Hamilton–Jacobi equation offers a natural and useful area of concentration since it is of interest for its own sake, occurring in numerous classical applications, and since some limits must be imposed on the scope of a book of this nature.

The Hamilton–Jacobi equation arose early in the study of the calculus of variations since the local minimum of a variational problem with variable endpoints satisfies this equation, when viewed as a function of the endpoints. Jacobi then turned to the problem of finding the extremal curves, or characteristics, from the solution of the Hamilton–Jacobi equation, thus using the theory of PDEs to find solutions of certain systems of ordinary differential equations (ODEs) called Hamiltonian systems.

Around the turn of the century, A. R. Forsyth, in his monumental six-volume work on differential equations (see [111]), considered problems that could be labeled global solutions. This created no special interest and was largely ignored. The modern study of global solutions of nonlinear equations probably began in 1950–1951 with the papers of Hopf [143] and Cole [67] analyzing global solutions of Burger's equation. Since that time, extensive work has been done in this area. Hyperbolic systems and shocks have been a major area of study, particularly by P. D. Lax. The present book considers only the single first-order equation. Oleinik's beautiful paper [203] of 1957 analyzed the one-dimensional Cauchy problem, obtaining existence and uniqueness theorems by various methods. The variational method was introduced into the study of global solutions in 1964 by Conway and Hopf [73]. Since then, many different methods have been used by Hopf, Fleming, Douglis, Kruzhov, Aizawa and Kikuchi, Kuznetsov and Rozhdestvenskii, Elliott and Kalton, this author, and others. Very general existence and uniqueness theorems for the Cauchy problem were obtained by Douglis [87], and general existence theorems for boundary value problems have been obtained by this author.

Originally, this book was intended to be self-contained. This goal soon proved unfeasible because of the volume of background material necessary to understand all of the works just mentioned. Most of the author's work required a minimum of background, and the proofs actually presented in this monograph may be understood by the reader with little formal background in pure mathematics. However, many theorems of other authors are presented without proof and a thorough study of their proofs by the interested reader will require some background in topology, analysis, convex functions, and stochastic processes. Some background in ODEs and PDEs is also desirable. Several textbooks in these areas are included among the references, although no attempt at completeness is made. One concept from Douglis has been used without definition: the concept of semiconcavity. A function $f: \mathbb{R}^n \to \mathbb{R}$ is said to be semiconcave with constant K if, for all x and h in $\mathbb{R}^n$

$$f(x + h) + f(x - h) - 2f(x) \leq K|h|^2.$$

This definition will be used throughout most of chapter III, including the presentation of Douglis' theorem, but will be modified slightly in the discussion of the work of Feltus.

Acknowledgments

The author is deeply indebted to E. D. Conway for his extensive aid in the early phases of the research culminating in the existence chapter of this book. Appreciation is also due to S. I. Rosencrans, whose excellent course in PDEs at Tulane University furnished most of the material for chapter I, although the present author assumes responsibility for errors or lack of clarity introduced in the attempt to condense this material.

It is a pleasure to thank my colleagues at Montclair State College, W. A. Chai, T. E. Williamson, and P. W. Zipse, for valuable suggestions and aid in the preparation of this book. In particular, Professor Zipse was extremely helpful in developing the sections on the compatibility condition and the maximality of the variational solution. I would also like to express my appreciation to Montclair State College itself for a lightening of teaching load to allow preparation of this monograph and for a grant. For the typing of the manuscript, I would like to thank Mrs. Vicki Berutti. Finally, I thank Dr. Richard Bellman for the original suggestion that this book be written.

Introduction

The central problem of this book is to solve the Hamilton–Jacobi equation

$$u_t + H(t, x, u_x) = 0,$$

where t is a real variable (time) and x is an $\mathbb{R}^n$-valued variable (space). Here u_t denotes the partial derivative with respect to time and u_x the gradient in the space variables. H is a given function of $2n + 1$ real variables and a real solution $u(t, x)$ is sought.

It will be required that u satisfy given initial or boundary conditions. In the most general case, the "boundary" will be any closed set B in space–time and the initial conditions will be given in the form of a continuous real function f on B. The requirement then is that

$$u \,|\, B = f.$$

This problem will be referred to as the boundary value problem (H, f).

If B is contained in the hyperplane $t = b$, for some real constant b, this problem will be called a Cauchy problem. When B is a cylinder, $\mathbb{R} \times X$, with X fixed in space ($\mathbb{R}^n$), we have what is generally referred to as a boundary value problem, while if B consists of such a cylinder together with part of the hyperplane $t = b$, this is commonly alluded to as a "mixed" problem.

Because of the distinguished use of the time variable, the Hamilton–Jacobi equation is essentially an evolution equation and the Cauchy problem is more easily handled and more completely understood than the general

1

boundary value problem. This is true of the classical methods, as becomes apparent from characteristic theory, and extends to the more modern methods, where the Cauchy problem was the first problem solved, and has desirable uniqueness properties.

The important classical methods are local in nature and the domain of definition of the solution is generally severely restricted by any nonlinearity in H. These methods also require an excessive degree of smoothness, both of H and of f, hence B. It was for these reasons that global methods were developed. Global methods produce a solution defined in all of a given domain, usually $t > b$ and all x. In order to do this, they must necessarily relax the smoothness requirements on the solution, requiring an "almost everywhere" solution.

To avoid some of the cumbersome notation inherent in the classical treatments of partial differential equations we will employ a somewhat abbreviated notation for derivatives. For any point x of $\mathbb{R}^n$, x^j will denote the jth component of x. Now if $g: \mathbb{R}^n \to \mathbb{R}^m$ is differentiable and $y = g(x)$, the derivative of g may be identified with the $m \times n$ matrix of partial derivatives

$$Dg(x) = |a_{ij}| = |\partial y^i / \partial x^j|,$$

whose column vectors represent the partial derivative of y with respect to x^j, and whose row covectors represent the differential of y^i. This matrix will be denoted simply by

$$y_x = Dg(x).$$

Now if $h: \mathbb{R}^m \to \mathbb{R}^p$ is also differentiable, then the chain rule, in terms of matrix multiplication, will be denoted simply by

$$z_x = z_y y_x,$$

where $z = h(y)$. Thus if P is a point of $\mathbb{R}^m$, the expression Py_x denotes matrix multiplication, where P is thought of as a row vector. If Q is another point of $\mathbb{R}^m$, PQ denotes matrix multiplication where Q is now considered as a column vector, that is, PQ is simply the Euclidean inner product of P and Q. Admittedly this notation sometimes makes it difficult to remember which quantities are scalars, which are vectors, and which are matrices, but the increased simplicity of notation amply compensates for this difficulty.

One final notational point: the letter k will always stand for a positive integer and when used as a subscript denotes the kth term of a sequence. Thus the phrase "t_k remains bounded" will be understood to mean that $\{t_k : k = 1, 2, \ldots\}$ is a sequence and there is a positive constant M such that for every k, $-M < t_k < M$.

I Classical Methods

PREFATORY REMARKS

It is assumed that the reader is familiar with the elementary classical methods of partial differential equations. This chapter, however, will review these techniques briefly. No systematic development of classical theories is attempted, but rather their application to the Hamilton–Jacobi equation is treated together with their limitations. A collection of simple examples and problems will be presented. These will be treated by the various classical methods and will serve as a comparison for the results of the global techniques studied later in the book. Perhaps the most important point to be considered in this chapter is the impossibility of extending smooth local solutions to global solutions in general.

The common practice of denoting u_x by p, $\partial u/\partial x^j$ by p_j, and u_t by p_0, will be followed in this chapter.

Readers desiring more information about classical methods for first-order equations may consult any of several fine texts. Particularly recommended is Courant and Hilbert [77]. Other popular texts include Bers *et al.* [39], Chester [60], Garabedian [114], John [151], and Lax [176]. The more advanced researcher is referred to Ames [7, 9], Bluman and Cole [43], Carroll [53], and Mizohata [196]. Several other excellent texts are included in the references.

1. EXACT EQUATIONS AND DIRECT INTEGRATION

In some "degenerate" cases, partial differential equations may be reduced immediately to simple integration. Although such cases are trivial to solve and relatively rare in practice, an engineer or scientist may save extensive unnecessary labor by recognizing these equations.

For example, the n-dimensional Hamilton–Jacobi equation

$$u_t + H(t, x) = 0,$$

subject to the constraint

$$u(0, x) = f(x),$$

is immediately seen to have the unique solution

$$u(t, x) = f(x) - \int_0^t H(\tau, x) \, d\tau.$$

The slightly more general equation

$$u_t + H(t, x, u) = 0,$$

with the same constraint, is handled simply by treating it as an ordinary differential equation in t with "parameters" x:

$$u'(t) = -H(t, x, u(t)),$$

$$u(0) = f(x).$$

In general, any first-order equation $F(x, u, p) = 0$, in which p_j is missing, may be considered an equation in the x^i, $i \neq j$, with x^j a parameter, thus reducing the number of independent variables by one.

Another case where direct integration is useful is the exact equation, although the concept is not as common in partial differential equations as it is in ordinary differential equations. Consider the quasilinear PDE

$$M(x, y, u)u_x = N(x, y, u)u_y,$$

where M and N are continuously differentiable and satisfy the exactness condition

$$M_x = N_y.$$

In this case, u may be given implicitly by

$$\Phi(x, y, u) = 0,$$

where $M = \Phi_y$ and $N = \Phi_x$. To determine the "integral" Φ, set

$$\Phi = \int M \, dy + g(x, u).$$

Then since $\Phi_x = N$,

$$\int M_x \, dy + g_x(x, u) = N.$$

Solving for $g_x(x, u)$ we have

$$\Phi = \int M \, dy + \int g_x(x, u) \, dx + h(u).$$

For any h such that Φ_u does not vanish, the resulting u is easily seen to be a solution of the given PDE.

Example 1.1   Consider the equation in one space variable $(x \in \mathbb{R})$

$$xu_t = tuu_x.$$

Letting $M = x$, $N = tu$, the equation is exact since $M_t = 0 = N_x$. Then

$$\Phi = \int x \, dx + g(t, u) = \tfrac{1}{2}x^2 + g(t, u).$$

To determine g we use

$$0 + g_t(t, u) = tu.$$

Thus

$$\Phi(t, x, u) = \tfrac{1}{2}x^2 + \tfrac{1}{2}t^2u + h(u).$$

For a particular solution let

$$2h(u) = au + b.$$

Then

$$u(t, x) = -(x^2 + b)/(t^2 + a),$$

which is indeed a solution of the original PDE.   ∎

As in ordinary differential equations, it may be possible to find an integrating factor $\mu(x, y)$ so that

$$(\mu M)_x = (\mu N)_y.$$

If, for example, $(N_y - M_x)/M$ is a function of x alone, then

$$\mu(x) = \exp\left[\int \{(N_y - M_x)/M\} \, dx\right]$$

is an integrating factor.

Example 1.2   The Hamilton–Jacobi equation

$$u_t = xu_x$$

may be solved by this method. Let $M = 1$, $N = x$. Then

$$(N_x - M_t)/M = 1.$$

So

$$\mu(t) = \exp \int 1 \, dt = e^t$$

is an integrating factor, and the original equation becomes

$$e^t u_t = x e^t u_x.$$

This equation is exact, with integral

$$\Phi(t, x, u) = xe^t + h(u) = 0.$$

Letting $h(u) = -(au + b)$, the solution

$$u(t, x) = (xe^t - b)/a$$

results. Letting $h(u) = u^{1/3}$, one obtains the solution

$$u(t, x) = -x^3 e^{3t}. \quad \blacksquare$$

Since only very special equations are exact, since no general method exists for finding integrating factors, and since selection of the proper h to yield given initial values is often an impossible task, this method finds little use. It should also be noted that any implicit solution for u can, in general, be applied only locally since the implicit function theorem must be invoked.

2. THE GENERAL SOLUTION

The aim of the analyst is often to study the class of all solutions of a give PDE without reference to boundary conditions. This goal is often considered to be accomplished when one obtains the "general" solution. If the given PDE is

$$F(x, u, u_x) = 0,$$

with $x \in \mathbb{R}^n$ (n-independent variables), a general solution is a function $u(x;g)$, where g is an arbitrary real function of $n - 1$ variables and, for each fixed g, $u(\cdot\,; g)$ solves the PDE.

There is a vague secure feeling that possession of a general solution puts one in possession of "most" of the solutions of a given PDE, with only possibly a few "singular" solutions escaping. For most of the important equations of mathematical physics, this is true. However, the security is shaken if we note, as do Courant and Hilbert [77, p. 25], that if $F = GH$, any general solution of

$$G(x, u, u_x) = 0$$

is, by definition, a general solution of F. But clearly any solutions of

$$H(x, u, u_x) = 0,$$

which are not solutions of $G = 0$, are also solutions of $F = 0$ which are not included in this "general" solution.

With this warning in mind, we may proceed to show that many solutions are available from a general solution. For example, if there is a uniqueness theorem for a boundary value problem and the arbitrary function g represents the boundary data, then since every solution has boundary values, all solutions are contained in the general solution.

Example 2.1 For the simple equation $u_t = u_x$, a general solution is

$$u(t, x; g) = g(x + t).$$

Since there is only one solution taking on prescribed values for $t = 0$, if $v(t, x)$ is any solution, let

$$g(x) = v(0, x).$$

Then $v(t, x) = u(t, x; g)$, so v is indeed included in the general solution. The uniqueness assertion above will follow easily from the characteristic theory of section 8. ∎

The arbitrary function g may enter into the definition of the general solution in much more complicated manners than above. For example, a study of parameters may yield a general solution in which g enters in a very complicated manner and is practically unrecognizable in the final form of the solution. A study of the complete integral will demonstrate this more clearly. In these cases, determination of the correct g to yield desired boundary values may be impossible.

Example 2.2 The equation $x u_t = t u u_x$ of example 1.1 was shown to have a general solution $u(t, x; g)$ defined by

$$x^2 + t^2 u + g(u) = 0.$$

Suppose it is desired to find a solution with $u(0, x) = f(x)$. Inserting this requirement into the general solution, we obtain

$$x^2 + g \circ f(x) = 0.$$

This yields, on setting $z = f(x)$,

$$g(z) = -f^{-1}(z)^2,$$

provided that f is invertible. Even if f happens to be theoretically invertible (injective), one is left with the often onerous task of performing the inversion. Thus the general solution is of dubious value in this problem. ∎

Since no general method for obtaining the general solution is offered, the general solution may be considered, not a method of solution, but rather a goal of solution, to be effected by other means (complete integral, characteristics, etc.). Even when found, it often proves useless for the problem at hand. Thus little emphasis will be placed on the general solution in this monograph.

3. SEPARATION OF VARIABLES

One of the first methods introduced in elementary courses in partial differential equations is separation of variables. There are two important reasons for this. First, the method is very simple and sometimes even elegant. Second, many important problems of mathematical physics and engineering can be solved by this method. Unfortunately though, the method is extremely limited in scope of application.

Separation of variables consists simply of assuming a solution of the form

$$u(t, x) = g(t)h(x) \tag{3.1}$$

or of the form

$$u(t, x) = g(t) + h(x). \tag{3.2}$$

The given partial differential equation is then used to determine the form of g and h. A solution of this form is then made to satisfy the required boundary conditions if possible. In the case of a linear equation, the superposition principle allows a finite, or even infinite, number of solutions to be added in order to match the boundary data, leading to Fourier series solutions and similar methods. For this reason, fairly powerful results may be obtained by separation of variables in the linear case only.

Example 3.1 To solve the one-dimensional Hamilton–Jacobi equation

$$u_t + u_x{}^2 = 0, \qquad (3.3)$$

subject to

$$u(0, x) = x^2, \qquad (3.4)$$

assume a solution of the form (3.1). Then by (3.3),

$$g'h + (gh')^2 = 0$$

or

$$g'/g^2 = -(h')^2/h = c.$$

These equations have solutions

$$g(t) = a/(1 - act),$$
$$h(x) = -c(x - b)^2/4,$$

for some constants a, b, and c. Thus

$$u(t, x) = -ac(x - b)^2/4(1 - act).$$

Since (3.4) must hold, b and the product ac are determined and

$$u(t, x) = x^2/(1 + 4t), \qquad t \neq -\tfrac{1}{4}. \qquad \blacksquare$$

To get an idea of the applicability of this technique for $n = 1$, consider a function u as in (3.1). Then noting that

$$u_t = g'(t)h(x)$$

and

$$u_x = g(t)h'(x),$$

one obtains

$$u_t + (-g'h/gh')u_x = 0,$$
$$u_t + (-g'/g^2h')uu_x = 0,$$
$$u_t + (-g'h/g^k(h')^k)u_x{}^k = 0.$$

These equations give several types of Hamilton–Jacobi equation which may be solved by inspection. However, if a boundary condition, such as

$$u(0, x) = f(x)$$

is given, this may only be satisfied if

$$f(x) = g(0)h(x),$$

so the possible initial values f are extremely limited.

Similarly, (3.2) yields

$$u_t + (-g'/h')u_x = 0,$$

$$u_t + (-g'/(h')^k)u_x^k = 0,$$

and other similar equations, but again the possible initial values are extremely limited.

A slight generalization of this method consists of choosing an invertible transformation $(t, x) \mapsto (v, w)$ and then proceeding as before. This is equivalent to seeking a solution of form (3.1) or (3.2) except that g is now a function of v and h is now a function of w.

Example 3.2 Suppose it is desired to solve the one-dimensional problem

$$u_t + xu_x/t + t^2u_x^2 = 0,$$

$$u(t, 0) = 1 - t.$$

A solution may be obtained by setting $t = v$ and $x = vw$. Then these become

$$U_v + U_w^2 = 0,$$

$$U(v, 0) = 1 - v,$$

where $U(v, w) = u(t, x)$. By separation of variables, a solution is found to be

$$U(v, w) = 1 + w - v.$$

Thus the original problem has solution

$$u(t, x) = 1 + x/t - t, \qquad t \neq 0. \qquad \blacksquare$$

Exercises

3.1 What types of equation have solutions of the form

$$u(t, x) = g(t)h(x) + a(t) + b(x)?$$

For the Cauchy problem with data given for $t = 0$, what data $f(x)$ can be handled by the techniques of this section?

3.2 For general n, suppose

$$u(t, x) = g(t) \prod \{h_j(x^j) : 1 \leq j \leq n\}.$$

What type of equations now hold? Find a solution of this form for the problem

$$u_t + xu_x/t - (n + 1)u/t = 0,$$

$$u(t, t, \ldots, t) = at^{n+1}.$$

Is your solution unique among the equations obtainable by separation of variables?

4. CHARACTERISTICS AND INTEGRAL SURFACES

Before studying more general methods for solution of first-order equations, it is convenient to standardize the terminology involved. Consideration of the general first-order equation rather than the Hamilton–Jacobi equation actually simplifies the notation somewhat. The problem, then, is to solve

$$F(x, u, u_x) = 0, \tag{4.1}$$

subject to

$$u|B = f. \tag{4.2}$$

Since there are many different Euclidean spaces floating around and the terminology often becomes a little muddled, we will distinguish carefully among several related geometric objects. If u is a solution of (4.1)–(4.2), the set of points

$$\mathscr{S} = \{(x, z, p) \in \mathbb{R}^{2n+1} : z = u(x), p = u_x(x)\},$$

is called a *strip manifold*. The projection of $\mathscr{S}$ on $\mathbb{R}^{n+1}$ given by

$$\mathscr{I} = \{(x, z) \in \mathbb{R}^{n+1} : z = u(x)\}$$

will be called an *integral surface* and the projection of $\mathscr{I}$ into x-space is simply $\mathscr{D}$ the *domain* of u. The set B will always be an $(n - 1)$-dimensional manifold for the remainder of this chapter and will be called the *boundary* or *initial manifold*. The set

$$\mathscr{I}_0 = \{(x, z) \in \mathbb{R}^{n+1} : x \in B, z = f(x)\},$$

will be called an *initial surface* and

$$\mathscr{S}_0 = \{(x, z, p) \in \mathbb{R}^{2n+1} : (x, z) \in \mathscr{I}_0, p = u_x(x)\},$$

will be called an *initial strip manifold* or simply an *initial strip*. Any curve

$$s \mapsto (X(s), U(s), P(s)) \in \mathscr{S}$$

will be called a *characteristic strip* and its projection,

$$s \mapsto (X(s), U(s)) \in \mathscr{I},$$

a *characteristic curve*. Finally the projection on $\mathbb{R}^n$, $X(s)$, will be called a *base characteristic*. The reader is warned that this terminology, while not uncommon, is by no means standard.

Notice now that the $(n + 1)$-vector, $V = (u_x(x), -1)$, is normal to the integral surface, so that if $(x, z, p) \in \mathscr{I}$, then the hyperplane defined by

$$(p, -1)(\xi - x, \zeta - z) = 0$$

is tangent to $\mathscr{I}$. Thus if $\mathscr{S}_0 \subset \mathscr{S}$ and $(x, f(x), p) \in \mathscr{S}_0$,

$$\mathscr{H} = \{(\xi, \zeta) \in \mathbb{R}^{n+1} : (p, -1)(\xi - x, \zeta - f(x)) = 0\} \tag{4.3}$$

is tangent to the initial surface $\mathscr{I}_0$ at $(x, f(x))$. This fact is referred to as the *initial strip condition* and may be used to determine the initial strip manifold from the initial surface.

To use the initial strip condition most easily we obtain an equivalent form. Toward this end let g be a coordinate patch on B. Then $h = g^{-1} : \mathcal{O} \subset \mathbb{R}^{n-1} \to \mathscr{D}$ is differentiable and $dh(r) = h_r$ has rank $n - 1$. The column vectors, $\partial h/\partial r^i$, $1 \le i \le n - 1$, span the tangent hyperplane to B at $x = h(r)$, so $(p, -1)(\partial h/\partial r^i, \partial f/\partial r^i)$ must vanish for each i. That is, if $\varphi = f \circ h$,

$$\varphi_r = p h_r. \tag{4.4}$$

Also

$$F(h(r), \varphi(r), p) = 0. \tag{4.5}$$

Equation (4.4) may be arrived at more directly by noticing that if $u(h(r)) = f(h(r))$ and $p = u_x(h(r))$, then (4.4) is simply the chain rule for this case. Together (4.4) and (4.5) determine the initial strip manifolds from the initial surfaces. Let a solution of (4.4)–(4.5) be denoted by $\rho(r)$. Then the corresponding initial strip is

$$\mathscr{S}_0 = (h(r), \varphi(r), \rho(r)).$$

Now if $F_p(h(r), \varphi(r), \rho(r))$ lies in the hyperplane tangent to B, that is, if

$$\det \begin{vmatrix} \partial h^1/\partial r^1 & \cdots & \partial h^1/\partial r^{n-1} & \partial F(h, \varphi, \rho)/\partial p^1 \\ \vdots & & \vdots & \vdots \\ \partial h^n/\partial r^1 & \cdots & \partial h^n/\partial r^{n-1} & \partial F(h, \varphi, \rho)/\partial p^n \end{vmatrix} = 0, \tag{4.6}$$

the manifold B is said to be *characteristic* at $x = h(r)$, although more properly the initial strip (h, φ, ρ) should be said to be characteristic. If B is not characteristic at $x = h(r)$, B is said to be *free* or *noncharacteristic* there. If B is free at every point, then it is called a free initial manifold, or a noncharacteristic initial manifold, and the problem (4.1)–(4.2) is called a noncharacteristic initial value problem. If B is given by $G(x) = 0$, where $dG \neq 0$, then the strip (h, φ, ρ) is characteristic if and only if

$$dG(h) F_p(h, \varphi, \rho) = 0.$$

Thus if B is free, F_p has component normal to B and we shall see that F_p can be used to determine the integral surface off from the initial surface, at least locally, as long as the functions involved are sufficiently smooth.

For further discussion of characteristic and free initial manifolds, and more general cases, see Courant and Hilbert [77], John [151], or Lax [176].

Exercises

4.1   For the linear equation

$$F(x, u, u_x) = a(x)u_x(x) - b(x)u(x) - c(x) = 0,$$

show that the boundary is free or characteristic independent of the data f so the term "characteristic manifold" is precise.

4.2   Extend the result of exercise 4.1 to the semilinear case

$$F(x, u, u_x) = a(x)u_x - c(x, u) = 0.$$

4.3   For the quasilinear case

$$F(x, u, u_x) = a(x, u)u_x - c(x, u) = 0,$$

show that the boundary is free or characteristic independent of $\rho(r)$ and thus that there is only one initial strip for any given initial surface.

4.4   For the Hamilton–Jacobi Cauchy problem

$$u_t + H(t, x, u_x) = 0,$$

$$u(0, x) = f(x),$$

show that the initial manifold is always free and that for any data f there is a unique initial strip as long as f is at least differentiable.

5. THE COMPLETE INTEGRAL

A complete integral of the first-order equation (4.1) in n variables is a solution depending upon n "independent" parameters. That is,

$$u(x) = z(x; q), \qquad q \in \mathbb{R}^n, \tag{5.1}$$

where, for each q, $F(x, z, z_x) = 0$ and the matrix

$$\begin{vmatrix} \partial z/\partial q^1 & \partial^2 z/\partial x^1\, \partial q^1 & \partial^2 z/\partial x^2\, \partial q^1 & \cdots & \partial^2 z/\partial x^n\, \partial q^1 \\ \partial z/\partial q^2 & \partial^2 z/\partial x^1\, \partial q^2 & \partial^2 z/\partial x^2\, \partial q^2 & \cdots & \partial^2 z/\partial x^n\, \partial q^2 \\ \vdots & \vdots & \vdots & & \vdots \\ \partial z/\partial q^n & \partial^2 z/\partial x^1\, \partial q^n & \partial^2 z/\partial x^2\, \partial q^n & \cdots & \partial^2 z/\partial x^n\, \partial q^n \end{vmatrix}$$

has rank n. The rank condition is to ensure that u is not actually a function of x and $n - 1$ or fewer parameters.

Given a complete integral, a general solution of (4.1) may be obtained by the method of envelopes. To accomplish this, let g be an arbitrary real function of $n - 1$ variables, and set

$$q^n = g(q^1, \ldots, q^{n-1}). \tag{5.2}$$

Any smooth envelope of any family of integral surfaces is again an integral surface. The envelope of the $n - 1$ parameter family of integral surfaces obtained by substituting (5.2) into (5.1) is thus a solution depending upon x and the arbitrary function g, hence a general solution. This envelope is obtained by equating all partial derivatives of

$$w(x; q^1, q^2, \ldots, q^{n-1}) = z(x; q^1, \ldots, q^{n-1}, g(q^1, \ldots, q^{n-1}))$$

with respect to $q^1, q^2, \ldots, q^{n-1}$ to zero. The resulting $n - 1$ equations, together with (5.2), yield n equations which, hopefully, may be solved for q in terms of x.

Of course this does not solve the boundary value problem. If boundary data f is given on an initial manifold B, perhaps q may be chosen so that (5.1) satisfies the initial data. If not, possibly a function g, as in (5.2) may be chosen so that the resulting solution satisfies the initial data. However, this is not possible in general, so we proceed as follows.

Suppose that $B = h(\mathcal{O})$ for simplicity and for $r \in \mathcal{O}$, define $\varphi(r) = f(h(r))$. Now define

$$G(r, q) = \varphi(r) - z(h(r), q). \tag{5.3}$$

If a strip manifold is to contain the initial strip defined by $h(r)$, $\varphi(r)$, and $z_x(h, q)$, where q is some function of r, G of (5.3) must vanish and the initial strip condition

$$G_r(r, q) = \varphi_r(r) - z_x h_r(r) = 0 \tag{5.4}$$

must hold. If these n equations can be solved for q as a function of r and the envelope condition

$$z_q q_r = 0,$$

can be solved for r as a function of x, then substitution of the resulting function q of x into (5.1) yields an integral surface containing the initial surface. The proof is straightforward and will be left as an exercise.

Let us turn now to the Hamilton–Jacobi equation. For the space-time independent Hamilton–Jacobi equation in n dimensions

$$u_t + H(u_x) = 0 \tag{5.5}$$

a complete integral is

$$u(t, x; d, q) = d + qx - tH(q), \tag{5.6}$$

where q is an n-vector and d is real. For any d and q, this is a smooth solution of (5.5) on all of $\mathbb{R} \times \mathbb{R}^n$. The parameters d and q may be chosen so that (5.6) satisfies any linear boundary data. If nonlinear boundary data is given, the method of envelopes must be resorted to.

Example 5.1   Consider the n-dimensional Hamilton–Jacobi equation

$$u_t + |u_x|^2 = 0. \tag{5.7}$$

Suppose for g in (5.2), one sets

$$d(q) = a + bq + cqq,$$

where a and c are real and b is an n-vector. Differentiating d,

$$\partial d / \partial q^j = b^j + 2cq^j, \qquad 1 \le j \le n,$$

so that differentiating u yields

$$b^j + 2cq^j + x^j - 2tq^j = 0,$$

and

$$q = (x + b)/2(t - c).$$

This yields

$$u(t, x) = a + (x + b)^2/4(t - c), \qquad t \ne c.$$

By a proper choice of a, b, and c, this solves the Cauchy problem for any quadratic initial data:

$$u(0, x) = f(x).$$

If f is linear, of course d and q may be chosen in (5.6) to yield a solution directly. ∎

Example 5.2 The one-dimensional Cauchy problem may be most easily handled by letting h be the identity map. The solution from (5.1), (5.3), (5.4) and the envelope condition becomes

$$u(t, x) = d + xq - tH(q),$$

$$f(r) = d + qr,$$

$$f'(r) = q,$$

$$(x - r - tH'(f'(r)))f''(r) = 0.$$

If f'' never vanishes, the last equation yields $x = r + tH'(f'(r))$. But even if f'' does vanish, this relation makes the envelope condition hold. As long as some smooth function $r(t, x)$ makes the envelope condition hold, the above system solves the initial value problem. Now let $H(q) = q^2$. Then if $f(r) = r$, f'' vanishes identically, but $r = x - 2t$ is a smooth solution of the envelope condition yielding the solution $u(t, x) = x - t$. If $f(r) = r^2$, one obtains $r = x/(1 + 4t)$, which leads to the solution $u(t, x) = x^2/(1 + 4t)$, $t \neq -\frac{1}{4}$, as in example 3.1. Finally, if $f(r) = \exp(r)$, the solution is given by setting $x = r - 2t \exp(r)$ but this cannot be solved explicitly for r. ∎

The complete integral may also yield solutions not included in the general solution. These are called singular integrals and may be obtained by finding the envelope of the n parameter family (5.1) with no reference to an arbitrary function. That is, z_q is set to zero and q eliminated from the resulting n equations. Using the fact that $F(x, z, z_x)$ vanishes identically in q, singular integrals may be obtained directly from F without finding a complete integral because

$$F_q = F_u z_q + F_p z_{xq} = 0,$$

where

$$z_{xq} = |a_{ij}| = |\partial^2 z/\partial x^i\, \partial q_j|.$$

Now, for a singular integral, z_q vanishes identically. Thus as long as the matrix z_{xq} is nonsingular, F_p must vanish. This yields n equations which, along with (4.1), may be used to determine u and q in terms of x.

Example 5.3 *Clairaut's Equation* Consider the equation in two variables, x and y,

$$u = xu_x + yu_y + h(u_x, u_y).$$

A complete integral is

$$z(x, y; a, b) = ax + by + h(a, b).$$

For a general solution, let $b = g(a)$, and eliminate a from

$$u = ax + g(a)y + h(a, g(a))$$

and

$$0 = x + g'y + h_a + h_b g'.$$

A singular integral is obtained by envelopes from

$$u = ax + by + h(a, b),$$
$$0 = x + h_a,$$

and

$$0 = y + h_b.$$

The singular integral may also be obtained directly from Clairaut's equation, where now $p = u_x$, $q = u_y$, by

$$u = xp + yq + h(p, q),$$
$$0 = x + h_p,$$

and

$$0 = y + h_q.$$

Clearly both methods yield the same singular integral. See Courant [77, p. 27] for this and other examples. ∎

Example 5.4 The Hamilton–Jacobi equation has no singular integrals since differentiation of

$$u_t + H(t, x, u_x) = 0$$

with respect to u_t yields 1, which clearly can never vanish. ∎

Exercises

5.1 If $z(x; q)$ is a complete integral for F and q and r are smoothly determined by (5.3), (5.4), and the envelope condition, prove that $u(x) = z(x; q \circ r(x))$ is indeed a solution of the boundary value problem for F and f.

5.2 In example 5.1, let $n = 1$, and for g in (5.2), set $d = q^3$. What is the domain of definition of the resulting solution u? Show that there are two smooth solutions in this domain and verify that they are solutions by direct computation.

5.3   In example 5.1 with $n = 1$, suppose that $d = g(q)$, where $g'' \leq 0$ everywhere. Show that, for $t > 0$ and any x, the function

$$v(q) = g'(q) + x - 2tq$$

is monotone. Therefore if v changes sign, Newton's method or *regula falsi* may be used to find a root of v. Substituting this root into (5.6) thus yields a numerical solution of (5.7) in one dimension. Find the initial values corresponding to this solution by setting $t = 0$ in the above equation and in (5.6).

5.4   Now in example 5.1 with $n = 1$, let the initial data be $f(r) = r^3$. As in exercise 5.2, find two solutions to this initial value problem. What is their domain of definition?

5.5   Formulate and prove a theorem yielding numerical solutions to the one-dimensional Cauchy problem of example 5.2.

6. THE CAUCHY–KOWALESKI THEOREM

The attacks reviewed so far yield few positive results in the area of actually solving a given PDE. Separation of variables produces only specialized results of specialized equations and no general method is available to determine a complete integral or a general solution. When solutions are obtainable, they are generally restricted to the boundary data that can be matched and to the locality of the boundary by the limitations of the implicit function theorem. The Cauchy–Kowaleski theorem yields the first existence guarantee for solutions and also provides both a method for constructing these solutions and insight into the consequences of the noncharacteristic property of the initial manifold. This method requires extreme smoothness (analyticity), however, and again produces a solution only locally.

The Cauchy–Kowaleski theorem states that the analytic, free initial value problem has an analytic solution, at least locally. The general case is proved most easily by transforming to a first-order quasilinear system with initial manifold $x^n = 0$. See the aforementioned references. We shall treat only the following special case and that only sketchily to indicate the method.

THEOREM 6.1   If F is analytic and $\partial F(x, u, p)/\partial p^n$ never vanishes, then the analytic Cauchy problem

$$F(x, u, u_x) = 0,$$
$$u(x^1, \ldots, x^{n-1}, 0) = f(x^1, \ldots, x^{n-1})$$

has an analytic solution, $u(x)$, in a neighborhood of the initial manifold $x^n = 0$.

Outline of proof First note that $\partial F/\partial p^n \neq 0$ is equivalent to the problem being noncharacteristic and implies that $F = 0$ may be solved for p^n:

$$p^n = G(x, u, p^1, \ldots, p^{n-1}). \tag{6.1}$$

Now, for a point y with $y^n = 0$, consider the formal power series:

$$u(x) = \sum C_\alpha (x - y)^\alpha,$$

where the summation is over all n-vectors α whose components are non-negative integers and z^α is defined to be the product of all the z^i raised to the corresponding power α^i. If there is a solution of this form, the coefficients may be obtained from Taylor's theorem as follows:

$$C_{0\cdots 0} = u(y) = f(y),$$
$$C_{10\cdots 0} = \partial u/\partial x^1 |_{x=y} = \partial f(y)/\partial y^1,$$
$$\vdots$$
$$C_{0\cdots 010} = \partial u/\partial x^{n-1} |_{x=y} = \partial f(y)/\partial y^{n-1},$$
$$C_{0\cdots 01} = \partial u/\partial x^n |_{x=y} = G(y, f(y), f_y(y)),$$
$$\vdots$$

In general, C_α is a polynomial with positive coefficients in the partial derivatives of G and f so is completely determined. On the other hand, if the series determined this way has a positive radius of convergence, u so defined can be shown to solve the problem as stated. To prove the convergence, the method of majorants is used. See Courant and Hilbert [77], John [151], or Lax [176].

Example 6.1 Let us reconsider the problem of example 3.1. That is, $u_t + u_x^2 = 0, u(0, x) = x^2$. By the Cauchy–Kowaleski theorem, this problem has an analytic solution near $(0, 0)$:

$$u(t, x) = \sum_{k=0}^{\infty} \sum_{i+j=k} C_{ij} t^i x^j.$$

Now upon differentiating $u_t + u_x^2 = 0$, one obtains

$$u_{tt} + 2u_x u_{xt} = 0,$$
$$u_{tx} + 2u_x u_{xx} = 0,$$

and similar equations for higher-order derivatives. Thus

$$C_{00} = u(0, 0) = x^2 = 0,$$
$$C_{01} = u_x(0, 0) = 2x = 0,$$
$$C_{10} = u_t(0, 0) = -4x^2 = 0,$$
$$C_{02} = u_{xx}(0, 0)/2 = 1,$$

and so on. Continuing in this manner, one obtains the nonzero coefficients: $C_{02} = 1, C_{12} = -4, C_{22} = 16, \ldots$ Thus $u(t, x) = x^2 - 4tx^2 + 16t^2x^2 - \cdots$ or

$$u(t, x) = x^2 \sum_{j=0}^{\infty} (-4t)^j$$

This is a geometric series which converges for $|t| < \frac{1}{4}$, and the sum is easily recognized to be

$$u(t, x) = x^2/(1 + 4t), \qquad -\tfrac{1}{4} < t < \tfrac{1}{4}.$$

Of course this expression may be extended to all $t \neq -\frac{1}{4}$, yielding the same solution as that obtained by separation of variables, but notice that the series method guarantees a solution only locally ($|t| < \frac{1}{4}$). ∎

Example 6.2   Now let us change the last problem to $u_t + u_x^2 = 0$, $u(t, 0) = -t$. Differentiating the equation yields the same relationships as in example 6.1, but now it is necessary to solve for u_x rather than for u_t. One obtains $u_t(t, 0) = -1$, and

$$u_x(t, 0) = \pm(-u_t(t, 0))^{1/2}$$

or $u_x(t, 0) = \pm 1$. This leads to two solutions, again determined by expanding about $(0, 0)$. They are

$$u(t, x) = -t - x,$$
$$v(t, x) = -t + x.$$

Therefore, this problem, although free, does not have a unique solution. ∎

The method of this section of course requires extreme smoothness. But even for analytic problems, the actual evaluation of the solution is so cumbersome as to be impractical except in extremely simple contrived cases. Thus the method is useful primarily to prove existence. Even here it is weak since the radius of convergence of a power series may be extremely small. (In example 6.1, it was $\frac{1}{4}$.) The following exercise is intended only for stout souls with unbounded patience.

Exercise

6.1 Does the problem $u_t + u_x^2 = 0$, $u(0, x) = \exp(x)$, have an analytic solution? Obtain the first few terms of the formal expansion of the solution around $(0, 0)$. Can you obtain a general term?

7. THE LEGENDRE TRANSFORM

A "nonvertical" hyperplane in $\mathbb{R}^n \times \mathbb{R}$ is a set

$$H = \{(y, z) : z - yq + r = 0\},$$

where the n-vector q and the real number r are called the plane coordinates of H. Now if

$$\mathscr{S} = \{(x, z) : u(x) - z = 0\}$$

is the graph of the smooth function u, the normal to $\mathscr{S}$ at (x, z) is the vector $(u_x(x), -1)$. Thus the tangent hyperplane is

$$H = \{(y, z) : (y - x)u_x - (z - u) = 0\},$$

which has plane coordinates $q = u_x$, $r = xu_x - u$. $\mathscr{S}$ may be described locally then as the envelope of its tangent hyperplanes, that is, by the plane coordinates (q, r) rather than by the "point coordinates" (x, u). If $u_x = q$ may be solved for x (inverted), then $\mathscr{S}$ may also be determined by

$$r = r(q) = xu_x - u.$$

When this is done, the transformation $(x, u) \mapsto (q, r)$ is called the Legendre transform. If x is indeed a function of q, then $r = xq - u$, so that

$$r_q = x_q q + x - u_x x_q = x$$

and

$$u = r_q q - r.$$

This allows the Legendre transform to be put in the symmetric form

$$u + r = xq,$$

$$q = u_x, \qquad x = r_q. \tag{7.1}$$

The purpose of the Legendre transform, from the analyst's point of view, is to transform (4.1) into

$$F(r_q, qr_q - r, q) = 0. \tag{7.2}$$

Hopefully (7.2) has a simpler form than (4.1). Note though that the initial condition (4.2) on a boundary determined by G,

$$u(x) = f(x) \quad \text{when} \quad G(x) = 0,$$

transforms to

$$qr_q - r = f(r_q) \quad \text{when} \quad G(r_q) = 0,$$

which may cause problems.

Example 7.1   The equation $xa(u_x) = c(u_x)$ transforms to the simple linear equation

$$a(q)r_q = c(q),$$

which may be solved by means of the characteristic theory of the next section.   ∎

The space–time independent Hamilton–Jacobi equation

$$u_t + H(u_x) = 0$$

transforms to

$$q_0 + H(q) = 0,$$

which is no longer a differential equation. Thus it appears that the Legendre transform is of little use here. For the general Hamilton–Jacobi equation

$$u_t + H(t, x, u_x) = 0,$$

the Legendre transformation may be applied to x and u only, leaving t unchanged. This produces

$$(qr_q - r)_t + H(t, r_q, q) = 0,$$

which is now a second-order equation of some complexity. In some cases this may be helpful but generally not.

However, it turns out that the Legendre transform is extremely useful here employed in a perhaps unexpected manner. The transform is applied to H itself, considered as a function of $p = u_x$ only. Thus we have

$$L(t, x, q) = pq - H(t, x, p), \tag{7.3}$$

$$q = H_p(t, x, p). \tag{7.4}$$

This important transformation is central to the classical study of the Hamilton–Jacobi equation. If H is twice continuously differentiable, and $H_{pp}(t, x, p)$ has rank n, then (7.4) defines p as a function of q, at least locally.

Differentiating (7.3) with respect to q, where $p = p(q)$, and considering (7.4), one obtains

$$p = L_q(t, x, q). \tag{7.5}$$

This means that H is the Legendre transform of L, justifying the alternate nomenclature "Legendre conjugate."

Exercise

7.1   In the one-dimensional case $H = H(q)$, show that if H' is surjective, continuous, and monotone, L is defined everywhere. Give a graphical interpretation and obtain L as an integral of $(H')^{-1}$.

8. CHARACTERISTIC THEORY

The complete integral yields a general solution of a first-order partial differential equation by simple algebraic manipulation, although this manipulation may be impossible to carry out in practice. However, the primary deficiency of the complete integral is that there is no general method to determine the complete integral itself. Thus we are led to the study of characteristics for the most general existence and uniqueness theorems of classical partial differential equations.

We consider again the initial value problem

$$F(x, u, u_x) = 0, \tag{8.1}$$

$$u|B = f. \tag{8.2}$$

where $F \in C^2 (\mathscr{D} \times \mathbb{R} \times \mathbb{R}^n)$ with $\mathscr{D}$ a domain in $\mathbb{R}^n$, $f \in C^2(B)$, and B is a C^2 manifold of dimension $n - 1$ in $\mathscr{D}$.

The idea of characteristic theory is to reduce this problem to a problem in ordinary differential equations. To this end, consider the characteristic equation

$$\frac{d}{ds} \begin{vmatrix} X(s) \\ U(s) \\ P(s) \end{vmatrix} = \begin{vmatrix} F_p(X, U, P) \\ F_p(X, U, P)P \\ -F_x(X, U, P) - PF_u(X, U, P) \end{vmatrix}. \tag{8.3}$$

Note that, if $(X, U, P): [0, T] \to \mathscr{D} \times \mathbb{R} \times \mathbb{R}^n$ satisfies this system, then

$$D_s F(X, U, P) = F_x X' + F_u U' + F_p P'$$

$$= F_x F_p + F_u F_p P - F_p(F_x + PF_u)$$

$$= 0,$$

so that F is constant along the "characteristic strip" (X, U, P). Of course the whole point is to ensure that F starts out equal to zero at $s = 0$, so that F will remain zero all along this strip. Then if $u = U$ and $u_x = P$, which strangely enough will turn out to be the case, (8.1) will be satisfied and (X, U, P) will indeed be a characteristic strip.

Since we seek only a local solution, let us assume for simplicity that $B = h(\mathcal{O})$, where $\mathcal{O}$ is a domain in $\mathbb{R}^{n-1}$ and h is a C^2 diffeomorphism from $\mathcal{O}$ onto B. As before, define φ to be the composition of f and h. Now consider the initial value problem (8.3) together with

$$(X, U, P)(0) = (h(r), \varphi(r), \rho(r)), \tag{8.4}$$

where, for $r \in \mathcal{O}$, $\rho(r)$ is obtained from (4.4) and (4.5). Because F is C^2, the right-hand side of (8.3) is C^1 in all variables and the system (8.3)–(8.4) has a C^1 solution $(X, U, P)(s; r)$ in a neighborhood of $s = 0$. For each C^1 solution ρ of (4.4)–(4.5), the solution (X, U, P) is C^1 in all variables including r.

Because h is a diffeomorphism, (4.4) defines a one-dimensional manifold of solutions ρ for each r. If F actually depends upon u_x, that is, the Jacobian $\det|\partial F/\partial \rho|$ is not equal to zero anywhere, then (4.5) defines an $(n - 1)$-dimensional solution space. Thus (4.4)–(4.5) together may have no solutions, one or several solutions, or a one-dimensional manifold of solutions for each r. If there is a unique solution, it is smooth by the implicit function theorem and all is fine. If there are multiple solutions which can be pieced together in various smooth manners, the original problem (8.1)–(8.2) has multiple solutions, while lack of a solution to (4.4)–(4.5) usually means that (8.1)–(8.2) has no solution.

Let us suppose now that $\rho = \rho(r)$ is a C^1 solution to (4.4)–(4.5), and that $(X, U, P)(s; r)$ is the unique smooth solution of (8.3)–(8.4) corresponding to ρ. Now if x is given, to solve the initial value problem (8.1)–(8.2), it is necessary to find $(s; r)$ such that $X(s; r) = x$. That is, the map $(s; r) \mapsto X(s; r)$ must be inverted. This may be done locally if B is free. That is, the determinant in (4.6) does not vanish at any point $h(r)$ of B. This follows from the fact that the determinant in (4.6) is the Jacobian of the above map at the point $(0; r)$. If this is the case, define

$$u(x) = U \circ X^{-1}(x) = U(s; r). \tag{8.5}$$

This leads to the following:

THEOREM 8.1   If F, f, and B are twice continuously differentiable, B is free, and ρ is a smooth solution of (4.4)–(4.5), then (8.5) solves (8.1)–(8.2) at least in a neighborhood of B. Furthermore, u of (8.5) is twice continuously differentiable.

Proof It has already been seen that X^{-1} exists locally, and that $U \circ X^{-1}$ is C^1. Now, along the strip $(X, U, P)(s; r)$, where r is fixed, F is constant. But by (8.4) and (4.5), the constant value of F must be zero. The proof will be completed by showing that $u_x(x) = P \circ X^{-1}(x)$, which is C^1.

Define the n-vector W to be

$$W(s; r) = U_r - PX_r.$$

Similarly define

$$V(s; r) = U_s - X_s P.$$

It happens that $V = 0$ identically by (8.3), so that

$$W_s = W_s - V_r$$

$$= X_s P_r - P_s X_r$$

$$= F_p P_r + (F_x + PF_u)X_r.$$

But F vanishes identically near B, so

$$0 = F_r = F_x X_r + F_u U_r + F_p P_r,$$

which leads to

$$W_s = -F_u(U_r - PX_r) = -F_u W.$$

Now, by (4.4), $W(0; r) = \varphi_r - \rho h_r = 0$. Thus for each r, $W(s; r)$ is a solution of

$$w'(s) = -F_u w(s),$$

$$w(0) = 0.$$

This is a linear ordinary differential equation with unique solution zero. That is, W vanishes identically.

Now, $U_r = u_x X_r$ by (8.5) and since $W = 0$, $U_r = PX_r$. Since V also vanishes, we have

$$(P - u_x)X_r = 0,$$

$$(P - u_x)X_s = 0.$$

This system of n linear equations in n "unknowns," $P^j - \partial u/\partial x^j$, has non-singular matrix near B by the fact that B is free. Because of this, the only solution is zero, and $P = u_x$. ∎

THEOREM 8.2   Suppose that u is a C^2 solution of (8.1)–(8.2) in a neighborhood of B, that F, f, and B are also C^2 and that B is free with the initial strip defined by $\rho(r) = u_x(h(r))$. Then, subject to (8.3) and (8.4), u is given by (8.5), at least in some neighborhood of B.

Proof   See exercise 5.1.   ∎

COROLLARY   If F, f, and B are as in the theorem and (4.4)–(4.5) has a unique solution at each point of B and of course B is free, then (8.5) yields a unique C^2 solution of (8.1)–(8.2).   ∎

Let us return now to the Hamilton–Jacobi equation. Replace n by $n + 1$ in the preceding arguments, and let $t = x^{n+1}$. B is now an n-dimensional manifold in $\mathbb{R} \times \mathbb{R}^n$, $\mathcal{O}$ a domain in $\mathbb{R}^n$, and $B = h(\mathcal{O})$ as before. Now if u_t is denoted by p_0, and $p = (p_1, \ldots, p_n)$, then

$$F = p_0 + H(t, x, p),$$

and B is free at a point $h(r)$ if and only if

$$\det \begin{vmatrix} \partial h^0/\partial r^1 & \cdots & \partial h^0/\partial r^n & 1 \\ \partial h^1/\partial r^1 & \cdots & \partial h^1/\partial r^n & \partial H(h, p)/\partial p_1 \\ \vdots & & \vdots & \vdots \\ \partial h^n/\partial r^1 & \cdots & \partial h^n/\partial r^n & \partial H(h, p)/\partial p_n \end{vmatrix} \neq 0, \tag{8.6}$$

where

$$\partial \varphi/\partial r^j = \sum_{i=0}^{n} \rho_i(r)\, \partial h^i(r)/\partial r^j, \qquad 1 \le j \le n, \tag{8.7}$$

$$\rho_0(r) + H(h(r), \rho(r)) = 0. \tag{8.8}$$

The characteristic equations uncouple to some extent and become

$$X^0(s; r) = s + h^0(r), \tag{8.9}$$

$$X'(s; r) = H_p(X^0, X, P), \tag{8.10}$$

$$P'(s; r) = -H_x(X^0, X, P), \tag{8.11}$$

$$U'(s; r) = H_p P - H, \tag{8.12}$$

subject to

$$X(0; r) = (h^1(r), \ldots, h^n(r)),$$

$$P(0; r) = \rho(r),$$

$$U(0; r) = \varphi(r).$$

Note that (8.9) is already solved, and $P_0(s; r)$ is not needed, although the additional equation $P_0' = -H_t$ with $P_0(0) = \rho_0$ is useful in some other contexts. Equations (8.10) and (8.11) must be solved together and are called a Hamiltonian system. Equation (8.12) may then be solved by simple integration. Letting L be the Legendre conjugate of H, we have

$$L(t, x, q) = H_p(t, x, p)p - H(t, x, p),$$

where

$$q = H_p(t, x, p),$$

and the following equation holds:

$$U'(s; r) = L(X^0, X, X'). \tag{8.13}$$

Thus once the base characteristic $X(s; r)$ is known, U is obtained by integrating L along the characteristic.

Note finally that if we have a Cauchy problem with data given on all of the hyperplane $t = 0$, then $h^0(r)$ is identically zero and the fact that h is a diffeomorphism forces (8.6) to hold automatically. That is, the initial manifold is always free. Since h^0 is always zero, (8.7) does not contain the variable ρ_0, and always has a unique solution $\rho = (\rho^1, \ldots, \rho^n)$. ρ_0 may then be determined uniquely by (8.8). This substantiates the earlier remarks about the relative simplicity of the Cauchy problem by showing that the C^2 Cauchy problem always has a unique C^2 solution locally.

Example 8.1 Let $n = 1$, $\mathcal{O} = \mathbb{R}$, $h(r) = (0, r)$, $B = \{0\} \times \mathbb{R}$. This is therefore a one-dimensional Cauchy problem. Equation (8.6) becomes

$$\det \begin{vmatrix} 0 & 1 \\ 1 & H_p(0, r, \rho(r)) \end{vmatrix} = -1 \neq 0,$$

and (8.7) and (8.8) are

$$\varphi'(r) = \rho(r),$$

$$\rho_0(r) = -H(0, r, \varphi'(r)).$$

Equation (8.9) yields $X^0(s; r) = s$. Suppose now that $H(t, x, p) = p^2$. Then (8.11) becomes $P(s; r) = \varphi'(r)$ and (8.10) is

$$X(s; r) = r + 2s\varphi'(r).$$

Finally (8.12) yields

$$U(s; r) = \varphi(r) + s(\varphi'(r))^2.$$

Consider now the data $f(x) = x^2$, so that $\varphi(r) = r^2$. Then

$$t = X^0 = s,$$

$$x = r + 4tr,$$

$$U = r^2 + 4sr^2.$$

Solving for s and r yields

$$s = t,$$

$$r = x/(1 + 4t).$$

Thus

$$u(t, x) = x^2/(1 + 4t).$$

This solution is defined everwhere except when $t = -\frac{1}{4}$, and is the same solution obtained in examples 3.1 and 6.1. ∎

Example 8.2   Now with $n = 1$, $\mathcal{O} = \mathbb{R}$, let $h(r) = (r, 0)$, $B = \mathbb{R} \times \{0\}$, so that we have a classical boundary value problem rather than a Cauchy problem. Let $H(t, x, p) = H(p)$, independent of t and x. Then (8.7) and (8.8) are

$$\varphi'(r) = \rho_0(r),$$
$$H(\rho(r)) = -\varphi'(r).$$
$$(8.14)$$

The initial manifold is now noncharacteristic only if $H'(\rho)$ never vanishes. Consider now the data $f(t) = -t$. Then the cardinality $\aleph$ of the solution set of (8.14) for various Hamiltonians H is tabulated here:

$H(p)$:	$-p^2$	$\exp(p)$	p^2	$p^3 - p + 1$	$\sin(p)$	1
$\aleph$:	0	1	2	3	$\aleph_0$	c

Following fairly standard terminology, $\aleph_0$ is the cardinality of the natural numbers, and c that of the real line. It is immediately apparent that the only Hamiltonian of those listed which has a guaranteed unique solution is $H(p) = \exp(p)$. Let us consider what happens in the case of the Hamiltonian $H(p) = p^2$, where (8.14) has two solutions. These two smooth solutions are

$$\rho(r) = -1,$$

and

$$\rho(r) = 1.$$

These lead to two solutions of the boundary value problem:

$$u(t, x) = -x - t,$$
$$v(t, x) = x - t,$$

as were obtained previously by series methods in example 6.2. ∎

Example 8.3   With the situation of example 8.2, with $f(t) = 0$, consider the Hamiltonian $H(p) = p^2$. Then (8.14) has $\rho = 0$ as its only solution, so that B is characteristic and, by (8.10), X is constant along any characteristic. That is, X is always zero, and the characteristics do not leave the initial manifold. Because of this difficulty, the method of characteristics fails here. However the problem (H, f) obviously has the solution $u(t, x) = 0$. ∎

Example 8.4   Again in the situation of example 8.2 with $f(t) = -t$ let the Hamiltonian by $H(p) = p^3/3 - p^2/2 + 1$. Then (8.14) has as solutions, $\rho = 0$ and $\rho = \frac{3}{2}$. For $\rho(r) = 0$, the problem is characteristic but for $\rho(r) = \frac{3}{2}$, the problem is free, showing that the property of being free or characteristic is indeed a property of initial strips rather than of initial manifolds. ∎

Example 8.5   In the Cauchy problem of example 8.1, change the data to

$$f(x) = \sin(x/\epsilon), \qquad \epsilon > 0.$$

Then $t = s$, $x = r + 2s\epsilon^{-1}\cos(r/\epsilon)$ and

$$U(s; r) = \sin(r/\epsilon) + s\epsilon^{-2}\cos^2(r/\epsilon).$$

Consider now the characteristic starting at $r = 0$ and that from $r = \pi\epsilon/2$. Along the first $X(s) = 2s/\epsilon$, while along the second $X(s) = \pi\epsilon/2$. These intersect at $(t, x) = (\pi\epsilon^2/4, \pi\epsilon/2)$. Thus at this point of intersection, $u(t, x)$ must be given by $U(t, 0) = \pi/4$, but also by $U(t, \pi\epsilon/2) = 1$. Obviously this is impossible, so we conclude that there is no C^2 solution u whose domain includes the point (t, x). In fact, it can be shown that no C^2 solution may be extended for all x past $t = \epsilon^2/2$. ∎

The problem encountered in the last example is not an isolated occurrence but is in the nature of things. The solution guaranteed by characteristic theory is defined only where $(s; r) \mapsto X$ is invertible, that is, as long as the characteristics do not intersect. Where two characteristics intersect, data may be thought of as being brought up from the boundary along both, so the value of the solution is overdetermined. Any nonlinearity in the Hamiltonian leads to crossing of characteristics except with very special data. In general, therefore, solutions may exist only in an extremely restricted neighborhood of the initial manifold.

Exercises

8.1   Prove theorem 8.2 by considering $U = u$ and $P = u_x$ along a curve $X(s; r)$, where X has the characteristic slope

$$X' = F_p(X, U, P).$$

8.2   In example 8.5, show that if $t > \epsilon^2/2$, there are distinct characteristics which intersect at a point (τ, x) with $\tau < t$.

9. ELEMENTARY TRANSFORMATIONS

Sometimes very simple transformations are helpful in solving partial differential equations. A sample is provided by example 3.2 and another by the Legendre transformation. Elementary changes of variable are often examined with a view toward reduction to an equation amenable to separation of variables, exact methods, series methods, etc. In this section a few more basic transformations will be presented and the generality of the Hamilton–Jacobi equation made more clear.

Actually, some of the most powerful methods of PDE are extensively developed transform methods. The contact transformations and the similarity methods are among these methods. For this section, though, we shall stick to very simple transformations which prove extremely useful, at least locally. Ames [7, 9] systematically develops a wide range of transform methods.

Transformations may be applied to the "independent variables" (x), the "dependent variable" (u), the equation itself (F), or any combination of these. For a starter we consider a simple change of variables which reduces a PDE to an ODE.

Example 9.1   The equation in two independent variables

$$tu_t + xu_x = 2txu, \qquad u(0, x) = c,$$

is transformed by $v = tx$ into the ODE

$$u_v = u, \qquad u(0) = c,$$

which has solution

$$u(v) = ce^v.$$

Thus the original problem has solution

$$u(t, x) = ce^{tx}.$$

More generally, the PDE

$$F(tx, u, u_t/x, u_x/t) = 0$$

is transformed by $v = tx$ into the ODE

$$F(v, u, u_v, u_v) = 0.$$

However, initial conditions may be a real problem since the single variable v can take on only a single value on any manifold $tx = $ constant.   ∎

Example 9.2   The problem

$$(xu_x + tu_t)(t^2 + x^2)^{-1/2} + (xu_t - tu_x)^2 = 0,$$

$$u(0, x) = -x, \qquad x \geq 0,$$

naturally suggests polar coordinates. Sure enough, the transformation

$$t = r \sin \theta,$$

$$x = r \cos \theta,$$

transforms this problem to

$$U_r + U_\theta^2 = 0,$$

$$U(r, 0) = -r, \qquad r \geq 0.$$

This problem has solutions $U(r, \theta) = -r \pm \theta$ among others. Thus the original problem has solutions

$$u(t, x) = -(t^2 + x^2)^{1/2} \pm \tan^{-1}(t/x). \qquad \blacksquare$$

Next we consider the addition of an exact integrand to our problem. Specifically, suppose that u solves the Hamilton–Jacobi problem with Hamiltonian H and boundary data f on boundary B. If $w(t, x)$ is differentiable in a neighborhood of B, obviously $v = u + w$ solves the problem

$$v_t + \mathscr{H}(t, x, v_x) = 0,$$

$$v|B = f + w,$$

where

$$\mathscr{H}(t, x, p) = H(t, x, p - w_x(t, x)) - w_t(t, x).$$

Note also that if H has Legendre conjugate L, $\mathscr{H}$ has Legendre conjugate $\mathscr{L}$, where

$$\mathscr{L}(t, x, q) = L(t, x, q) + qw_x + w_t.$$

Example 9.3   Consider the problem

$$(v_x + t)^2 + v_t + x = 0,$$

$$v(0, x) = x^2,$$

Letting $w(t, x) = -tx$, and $v = u + w$, the problem becomes

$$u_t + u_x^2 = 0,$$

$$u(0, x) = x^2,$$

which has unique solution $u = x^2/(1 + 4t)$ for t near zero. Thus the original problem has unique solution

$$v(t, x) = x^2/(1 + 4t) - tx. \quad \blacksquare$$

The remainder of this section traces a series of transformations which may, in theory, be applied to any first-order equation to reduce it to various simpler or symmetric forms including Hamilton–Jacobi equations. The object of our attack is the general first-order boundary value problem (4.1)–(4.2). If this problem has a solution u, it may always be represented implicitly as

$$V(x, u) = 0, \qquad V_u \neq 0. \tag{9.1}$$

We will search directly for V as a function of $n + 1$ variables (x, u), rather than for u. Differentiating V, one sees that

$$\partial V/\partial x^i + V_u \, \partial u/\partial x^i = 0,$$

or

$$p_i = \partial u/\partial x^i$$

$$= -(\partial V/\partial x^i)/V_u, \qquad 1 \leq i \leq n.$$

Letting $X_i = x^i$, $1 \leq i \leq n$, $X_{n+1} = u$, and $P_i = \partial V/\partial X_i$, we have

$$F(x, u, p) = F(X_1, \ldots, X_{n+1}, -P_1/P_{n+1}, \ldots, -P_n/P_{n+1})$$

$$= G(X, P)$$

$$= 0. \tag{9.2}$$

When x is a point of B, $u(x) = f(x)$, so the boundary condition becomes

$$V(x, f(x)) = 0, \qquad x \in B. \tag{9.3}$$

Thus the original problem (4.1)–(4.2) has been transformed into the problem (9.2)–(9.3) in V, in which V does not occur explicitly, but one more independent variable is needed. Once V has been found, (9.1) yields u. Note that if $V(X; \alpha)$ is a solution of (9.2), where α represents n "independent parameters," $V(X; \alpha) + r$ is also a solution because V is missing in (9.2). Thus $V(X; \alpha) + r$ is a complete integral if r remains independent of the α's. So a complete integral of (9.2) really needs only n parameters, which would be expected since the original problem had only n independent variables. One of the chief advantages of eliminating u is that the characteristic equations (8.3) assume a symmetric form:

$$X'(s) = G_p(X, P),$$

$$P'(s) = -G_x(X, P). \tag{9.4}$$

After finding the solution (X, P) of this Hamiltonian system, V may be found by integrating

$$V'(s) = PG_p(X, P). \tag{9.5}$$

Example 9.4 The general quasilinear problem in n variables

$$a(x, u)u_x = c(x, u),$$

$$u(x) = f(x), \qquad x \in B,$$

transforms by this method into the homogeneous linear equation in $n + 1$ variables

$$A(X)V_x = 0,$$

where $A = (a, c)$, together with

$$V(x, f(x)) = 0. \qquad \blacksquare$$

Example 9.5 The nonlinear problem in two independent variables

$$u_y^2 + uu_x = xy,$$

$$u(0, y) = y,$$

becomes

$$V_y^2 = xyV_u^2 + uV_xV_u,$$

$$V(0, y, y) = 0,$$

which is still a complicated nonlinear equation but with V missing. This equation also lends itself to further transformation. $\blacksquare$

An equation with the dependent variable missing, $G(X, P) = 0$, can always be transformed into a Hamilton–Jacobi equation locally, as long as G_p never vanishes. For $G_p \neq 0$ implies that $G = 0$ may be solved for some P_i locally. Applying the transformation of (9.1) first, any first-order equation can be represented locally as a Hamilton–Jacobi equation. In this sense, we are studying very general first-order equations in this work.

Conversely, sometimes it is desirable to eliminate the specific dependence of the Hamiltonian on t. To accomplish this, all that is necessary is to introduce a new independent variable $x^{n+1} = t$ and $p_{n+1} = u_t$. Then letting

$$G(x, x^{n+1}, p, p_{n+1}) = p_{n+1} + H(x^{n+1}, x, p) = 0,$$

the equation has been written as $G(X, P) = 0$.

Example 9.6   The problem of example 9.5 yields upon solving for V_x the Hamilton–Jacobi equation

$$V_t + (ty V_u^2 - V_y^2)/u V_u = 0$$
$$V(0, y, y) = 0,$$

where $t = x$.   ∎

We complete this section with the reduction of any Hamilton–Jacobi equation to a quasilinear system. Rewriting the Hamilton–Jacobi equation as

$$u_t = -H(t, x, u_x),$$

we let p denote u_x as usual and $q = p_0 = u_t$. Then by a differentiation with respect to t and equating mixed partial derivatives we obtain the quasilinear system

$$u_t = q,$$
$$p_t = q_x, \tag{9.6}$$
$$q_t = -H_t(t, x, p) - H_p(t, x, p)q_x.$$

That is, if u is a twice continuously differentiable solution of the Hamilton–Jacobi equation, then (u, p, q) is a solution of (9.6), assuming also that H is at least differentiable. However, (9.6) generally has more solutions than the original equation. The corresponding Cauchy problems turn out to be equivalent though. If u satisfies the Hamilton–Jacobi equation in addition to the initial condition

$$u(0, x) = f(x),$$

(u, p, q) satisfies (9.6) and

$$u(0, x) = f(x),$$
$$p(0, x) = f_x(x). \tag{9.7}$$
$$q(0, x) = -H(0, x, f_x(x)).$$

Conversely suppose that (u, p, q) is a solution of (9.6)–(9.7). Then $u_{tx} = u_{xt} = p_t$, so integration yields

$$u_x(t, x) = p(t, x) + g(x).$$

By (9.7), $g(x)$ vanishes, so $p = u_x$. Also

$$u_{tt} = q_t = -\partial H(t, x, p(t, x))/\partial t,$$

so that

$$u_t = q = -H(t, x, p) + h(x).$$

Again (9.7) implies that h vanishes, so $u_t = -H$ and u satisfies the Hamilton–Jacobi equation.

In the case of only one space variable ($n = 1$), this transformation is often handled a little differently. Ignoring q, we have the single equation

$$p_t = - dH(t, x, p(t, x))/dx \tag{9.8}$$

together with

$$p(0, x) = f'(x). \tag{9.9}$$

After p is found, u may be found by integration. However, (9.8) and (9.9) alone determine u only to within an arbitrary function of t, so the original problem must be checked to evaluate the correct arbitrary function.

Example 9.7 The problem of example 3.1,

$$u_t + u_x{}^2 = 0, \qquad u(0, x) = x^2,$$

transforms to the system

$$u_t = q,$$
$$p_t = q_x,$$
$$q_t = -2pq_x,$$

together with the initial conditions

$$u(0, x) = x^2,$$
$$p(0, x) = 2x,$$
$$q(0, x) = -4x^2.$$

Here the system is more difficult to solve than the original equation by direct methods, so that the original problem becomes a method of solving the system. ∎

Example 9.8 Consider the problem

$$v_t = -2vv_x, \qquad v(0, x) = 2x.$$

Upon recognizing $2vv_x$ as the derivative of v^2 with respect to x, let $v = u_x$, obtaining

$$u_{tx} = u_{xt} = - d(u_x^2)/dx,$$

$$u(0, x) = x^2 + c,$$

or

$$u_t = -u_x^2,$$

$$u(0, x) = x^2 + c.$$

This is immediately recognized as the problem of example 3.1, having solution

$$u(t, x) = x^2/(1 + 4t) + c.$$

Therefore, we have found a solution to our original quasilinear problem

$$v(t, x) = 2x/(1 + 4t).$$

Of course, reversing this problem, v is easily found by the theory of characteristics, yielding yet another method of solving this Hamilton–Jacobi equation for u. ∎

10. VARIATIONAL METHODS

The Hamilton–Jacobi equation is intimately connected with the calculus of variations, having its historical roots in that subject, as well as some very fundamental interrelations. For our study, since we are beginning with the Hamilton–Jacobi equation, the situation of section 8 is a good point of departure. That is, B is a smooth n-dimensional manifold, f is also smooth, and H is m-times continuously differentiable, where $m \geq 2$. The Lagrangian L is defined by (7.3) as the Legendre transform of H.

Consider now the problem of minimizing the functional

$$I(X) = f(s, y) + \int_s^t L(\tau, X(\tau), X'(\tau)) \, d\tau, \tag{10.1}$$

where $X: [s, t] \to \mathbb{R}^n$ and $X(s) = y$, $(s, y) \in B$, $X(t) = x$, $s < t$. The curves X will be restricted to some class $\mathscr{C}$ of admissible curves, which will all join (s, y) to (t, x). The class $\mathscr{C}$ may be any class containing at least the piecewise linear curves whose derivatives are suitably bounded. More precisely, $\mathscr{C}$ must contain at least the piecewise linear curves in some C^1 neighborhood.

Suppose for a fixed (t, x) and (s, y) there is an extremal, that is, an admissible curve X such that $I(X)$ is the minimum value of (10.1), and that X is interior to the class of admissible curves (in whatever function space is under consideration). Now if Y is another admissible curve and $\delta = Y - X$, define for small real r

$$g(r) = I(X + r\delta). \tag{10.2}$$

Setting $g'(0) = 0$ and integrating by parts, one sees that

$$\int_s^t \left\{ L_q(\tau, X, X') - \int_s^\tau L_x(\sigma, X, X')\, d\sigma \right\} \delta'(\tau)\, d\tau = 0.$$

By the Euler–Lagrange lemma (exercise 10.1), there is a constant vector c such that

$$L_q(\tau, X(\tau), X'(\tau)) = \int_s^\tau L_x(\sigma, X(\sigma), X'(\sigma))\, d\sigma + c. \tag{10.3}$$

This is the Euler equation in integrated form. If X is smooth (C^1), then the right-hand side of (10.3) is also smooth. It is easy to show that $L_{qq} = H_{pp}^{-1}$, where the arguments are related by (7.4) and (7.5). Thus the implicit function theorem shows that $X'(\tau)$ is a smooth function of τ and $X(\tau)$ locally, so that X is twice continuously differentiable. By iterating the argument, it may be seen that X is m-times continuously differentiable and (10.3) may be differentiated, yielding the Euler equation

$$dL_q(\tau, X, X')/d\tau = L_x(\tau, X, X'). \tag{10.4}$$

Solutions X of this famous differential equation yield suspected extremals, although it is now realized that little else is guaranteed in general.

Suppose now that we let $P(\tau) = L_q(\tau, X, X')$. Noticing that $L_x(t, x, q) = -H_x(t, x, p)$, (7.4) and (10.4) become

$$X'(\tau) = H_p(\tau, X, P), \tag{10.5}$$

$$P'(\tau) = -H_x(\tau, X, P). \tag{10.6}$$

This Hamiltonian system is exactly that of (8.10)–(8.11). Thus the variational extremals coincide with characteristics. By (8.13), U of equation (8.12) is just the variational minimum $I(X)$.

We would expect that, as (t, x) varies, the local variational minimum, $u(t, x) = I(X)$, does indeed satisfy the Hamilton–Jacobi equation. This may be shown easily by the method of geodesic coverings. See Young [272]. Alternately, if a family of extremals covers a neighborhood of B simply, the characteristic theory may be used to show that the Hamilton–Jacobi equation is satisfied. It might be added that H_{pp} is indeed of rank n in cases where the method of geodesic coverings applies, as is implied by the well-known Weierstrass condition that requires that $L(t, x, q)$ be locally convex in q along an extremal. Thus the classical variational theory, assuming sufficient smoothness and convexity as well as simple coverings of extremals, shows that local variational minimums solve the initial value problem (H, f) locally. Local variational maximums, with concavity in place of convexity, also solve the Hamilton–Jacobi initial value problem locally.

Much more information is available from the classical theory. For example, in cases of strict convexity, the classical theory of conjugate points shows that the smooth solution is impossible only along lower dimensional manifolds where certain determinants vanish. However this topic would lead us too far astray, so we refer the reader to the classical texts. Bliss [42] has a readable account of conjugate points in the plane in his first chapter.

Example 10.1   Consider the plane Cauchy problem: $u_t + u_x{}^2 = 0$, $u(0, x) = f(x)$. Here $H(t, x, p) = p^2$, and $L(t, x, q) = q^2/4$. The Euler equation is simply $L_q{}' = 0$, or $L_q = q/2$ is constant. That is, the characteristics are given by constant X' and thus are straight lines. It may be shown that these are indeed extremals, so a solution is given by

$$u(t, x) = \min I(X)$$

$$= \min_y \left\{ f(y) + \int_0^t L((x - y)/t) \, d\tau \right\}$$

$$= \min_y \{ f(y) + (x - y)^2/4t \}.$$

This minimum may be found by simple single-variable calculus. Thus

$$2tf'(y) + (y - x) = 0.$$

The following table follows easily:

$f(y)$	$u(t, x)$
c	c
$ay + b$	$ax + b - a^2 t$
y^2	$x^2/(1 + 4t), \qquad t > -\tfrac{1}{4},$

agreeing with the results of previous examples. Now if $f(y)$ is $-y^2$, the above minimum does not exist for $t \geq \tfrac{1}{4}$. However, for $t < \tfrac{1}{4}$, one obtains $u(t, x) = x^2/(4t - 1)$. Replacing the minimum with a maximum, for $t > \tfrac{1}{4}$, the same formula for u holds, so this u is a solution for all $t \neq \tfrac{1}{4}$, but is a variational maximum for $t > \tfrac{1}{4}$ and a minimum for $t < \tfrac{1}{4}$. ∎

Exercise

10.1   Prove the following: Euler–Lagrange lemma. Let $g: [s, t] \to \mathbb{R}^n$ be an integrable function such that, for any piecewise linear function $\delta: [s, t] \to \mathbb{R}^n$ vanishing at s and t, with $\|\delta'\|_\infty$ sufficiently small,

$$\int_s^t g(\tau)\delta'(\tau)\, d\tau = 0.$$

Then g is a constant almost everwhere.

   Hint   For a unit vector e along the x^j axis, small ϵ,

$$\delta'(\tau) = \begin{cases} \epsilon e, & a < \tau < a + h, \\ -\epsilon e, & b - h < \tau < b, \\ 0, & \text{otherwise}, \end{cases}$$

where $s \le a < a + h < b - h < b \le t$. Let $\delta = \int \delta'$ and show that $g(a) = g(b)$ if a and b are Lebesgue points of g.

11. HAMILTON–JACOBI THEORY

The Hamiltonian system (10.5)–(10.6) is a system of ordinary differential equations to which an extensive theory may be applied. For this reason, the Hamiltonian system of characteristic equations is considered to yield the solution of the more difficult problem of solving the Hamilton–Jacobi equation. However, in many physical situations, the ODEs arise first and the Hamilton–Jacobi equation may be solved rather easily by some other method. Jacobi realized that the problem could be attacked in reverse and the PDE used to solve the ODEs.

First, let us show how "any" system of ODEs may be written as half of a Hamiltonian system. Suppose the given system is

$$x'(t) = F(t, x).$$

Then, letting

$$H(t, x, p) = pF(t, x), \tag{11.1}$$

one may set

$$x'(t) = H_p(t, x, p),$$
$$p'(t) = -H_x(t, x, p). \tag{11.2}$$

Since $H_p = F$, one sees that the original system has become the first half of this Hamiltonian system.

Suppose now that a complete integral

$$u = z(t, x; q) + a \tag{11.3}$$

of the Hamilton–Jacobi equation for the Hamiltonian of (11.1) is known. Here $q \in \mathbb{R}^n$ and $a \in \mathbb{R}$ are parameters. The general solution is obtained by letting $a = g(q)$ and taking envelopes as in section 5. For a given function g, the resulting u determines an integral surface. But for a fixed q and a, (11.3) also gives an integral surface. The intersection of these two integral surfaces is a curve, namely a characteristic, with corresponding base characteristic being a solution to (11.2). Letting g and its partial derivatives range over all possible functions, the general solution of (11.2) is obtained.

THEOREM 11.1   Let (11.3) represent a C^2 complete integral for the Hamilton–Jacobi equation corresponding to a C^1 Hamiltonian H. Assume further that $\det(z_{xq})$ never vanishes. Then the equations

$$z_q = \alpha \in \mathbb{R}^n, \tag{11.4}$$

$$z_x = p \tag{11.5}$$

determine implicitly a $2n$ parameter family

$$x = x(t; q, \alpha),$$

$$p = p(t; q, \alpha),$$

of solutions of (11.2).

Proof   Differentiating (11.4) with respect to t yields

$$z_{qt} + z_{qx} x_t = 0.$$

Then since

$$z_t + H(t, x, z_x) = 0,$$

it follows that

$$z_{tx} x_q + z_{tq} + H_x x_q + H_p p_q = 0.$$

But

$$z_{qt} = z_{tq} = -z_{qx} x_t$$

and

$$z_{tx} + H_x(t, x, z_x) = 0,$$

so

$$-z_{qx} x_t + H_p z_{xq} = 0.$$

Since z_{xq} is nonsingular,

$$x_t = H_p.$$

Similarly

$$z_{xt} + z_{xx}x_t = p_t$$

and

$$z_{tx} + H_x + H_p p_x = 0$$

imply that

$$p_t - z_{xx}x_t + H_x + H_p p_x = 0$$

or, since $x_t = H_p$ and $z_{xx} = p_x = 0$,

$$p_t = -H_x.$$

Thus x and p solve (11.2) as claimed. ∎

The same theory may be applied with slight variations to the general first-order PDE. See Chester [60, pp. 188–196].

Example 11.1   *The Two-Body Problem*   The problem of two point masses attracting each other according to a gravitational inverse square law leads to the vector equation

$$R''(t) = -kR/|R|^3,$$

where R is the time-varying position vector of one mass with respect to the other and k is a constant. It is easy to show that the motion is always planar, so that R may be represented by two coordinates $R = (x, y)$. See Pollard [218, pp. 16–19]. Although the above equation is a second-order equation, it may be reduced to Hamiltonian form by starting with the standard transformation

$$x' = p,$$
$$y' = q,$$
$$p' = -kx/(x^2 + y^2)^{3/2},$$
$$q' = -ky/(x^2 + y^2)^{3/2}. \tag{11.6}$$

As shown earlier, these four equations may be embedded in a Hamiltonian system of eight equations. However, it turns out that letting H be the total mechanical energy, we have

$$H(x, y, p, q) = (p^2 + q^2)/2 - k/(x^2 + y^2)^{1/2}.$$

The system (11.6) is already in Hamiltonian form:

$$x' = H_p,$$
$$y' = H_q,$$
$$p' = -H_x,$$
$$q' = -H_y.$$

The corresponding Hamilton–Jacobi equation is

$$u_t + (u_x^2 + u_y^2)/2 - k/(x^2 + y^2)^{1/2} = 0. \tag{11.7}$$

Introducing polar coordinates r and θ, we find (11.7) becomes

$$u_t + (u_r^2 + u_\theta^2/r^2)/2 - k/r = 0.$$

By separating variables and integrating, we obtain a complete integral

$$z(t, r, \theta; \alpha, \beta) = -\alpha t - \beta\theta - \int_\rho^r (2\alpha + 2k/\sigma - \beta^2/\sigma^2)^{1/2}\, d\sigma,$$

where $\rho = r(0)$. Setting $z_\alpha = -t_0, z_\beta = -\theta_0$ as in (11.4), we obtain the general solution of (11.6) in polar coordinates in terms of the four parameters α, β, t_0, and θ_0:

$$t - t_0 = -\int_\rho^r (2\alpha + 2k/\sigma - \beta^2/\sigma^2)^{-1/2}\, d\sigma,$$

$$\theta - \theta_0 = \beta \int_\rho^r \sigma^{-2}(2\alpha + 2k/\sigma - \beta^2/\sigma^2)^{-1/2}\, d\sigma. \tag{11.8}$$

Note that if t_0 is a fixed initial time α, β, ρ, and θ_0 may be thought of as the "parameters." Equation (11.8) is a relationship between r and θ and thus represents the trajectory or path of the motion. By a change of variables $\sigma \mapsto 1/\sigma$, (11.8) can be integrated, yielding

$$r = p/(1 - \epsilon^2 \sin(\theta - \theta_0)), \tag{11.9}$$

where

$$p = \beta^2/k,$$

and

$$\epsilon^4 = 1 + 2\alpha\beta^2/k^2.$$

Equation (11.9) is the polar form of the equation of a conic section, namely an ellipse, parabola, or hyperbola, depending upon whether $\epsilon < 1$, $\epsilon = 1$, or $\epsilon > 1$, which depends on turn upon whether H is negative, zero, or positive. (H may be shown to be constant rather simply.) ∎

For further examples, see Courant and Hilbert [77]. The Hamilton–Jacobi theory may also be attacked by the study of contact transformations, as will be shown in the next section.

12. CONTACT TRANSFORMATIONS

Transformations of (t, x, p) into (t, y, q) which leave the characteristic equations in Hamiltonian form, will be called contact or canonical transformations. It turns out that this is the case if and only if the old Hamiltonian H and the new Hamiltonian G are related by the equation

$$dF/dt = qy' - px' + (H - G), \tag{12.1}$$

where F is a differentiable function of t and either

   (i) x and y,
  (ii) x and q,
 (iii) p and y,

or

 (iv) p and q.

See Chester [60], Courant and Hilbert [77], or Goldstein [122]. The function F is called the generating function of the transformation and determines it completely. Naturally the aim is to choose F to make the resulting Hamiltonian system easier to solve than the original.

In case (i), since

$$dF/dt = F_x x' + F_y y' + F_t,$$

it is necessary only to take

$$p = -F_x,$$
$$q = F_y,$$

and

$$G = H - F_t$$

to obtain a contact transformation $(x, y) \mapsto (p, q)$, then solve for (y, q).

Case (ii) may be handled by applying a Legendre transform

$$F_1(t, x, y) = yq - F(t, x, q),$$

then proceeding as in case (i), to make dF_1/dt satisfy (12.1), obtaining

$$p = F_x,$$
$$y - F_q = 0,$$
$$G = H - F_t.$$

In case (iii), the relevant equations are

$$F_1(t, x, y) = xq - F(t, p, y),$$
$$x = F_p,$$
$$q = F_y,$$
$$G = H + F_t,$$

and in case (iv) a double Legendre transform yields

$$-F_1(t, x, y) = xp - yq + F(t, p, q),$$
$$x = -F_p,$$
$$y = F_q,$$
$$G = H + F_t.$$

Example 12.1   Let $F(t, x, q) = xq$. Then the transformation generated is

$$p = F_x = q,$$
$$y = F_q = x,$$
$$G = H,$$

or simply the identity transformation.   ∎

Example 12.2   A "point transformation" is a transformation $x \mapsto y$, independent of p. If $y = y(x)$ is any given function of x, the generating function $F(t, x, q) = qy(x)$ yields

$$p = F_x = qy_x,$$
$$y = F_q = y(x),$$
$$G = H,$$

or the given point transformation. Thus all point transformations are canonical.   ∎

Example 12.3 Let $F(t, x, y) = xy$. Then $p = -y$ and $q = x$, so the roles of x and p are reversed in y and q. In mechanics, x often represents (space) coordinates and p momenta. This transformation shows that the roles of coordinates and momenta are logically indistinguishable. ▌

Example 12.4 Suppose now that $z(t, x; q) + a$ is a complete integral of the Hamilton–Jacobi equation. Let $F(t, x, q) = z(t, x; q)$. The resulting transformation is $p = z_x$, $y = z_q$ and most important,

$$G = H(t, x, z_x) - z_t = 0.$$

That is, the transformed Hamiltonian system is

$$y' = G_q = 0,$$
$$q' = -G_y = 0,$$

which can be solved without difficulty. This is just one of the possible "canonical forms," but the preferred one if it can be achieved. Thus y and q are called canonical coordinates. Calling y a "parameter" α now since y is constant along a characteristic, it becomes apparent that this transformation is just that of (11.4)–(11.5), so that the contact transformation above accomplishes the Hamilton–Jacobi theory previously attained by taking envelopes.

In practice, the Hamilton–Jacobi theory is only useful when the Hamilton–Jacobi equation may be solved with relative ease. Otherwise the Hamiltonian system might better be attacked directly. If the transformed Hamiltonian $G(t, y, q)$ depends upon q only, then a complete integral is available. If G depends upon q and only one component y^i, then the PDE is separable and easily reduced to n ODEs which may be solved simply by reduction to quadratures (integration). See Goldstein [122, pp. 284–288]. This brief summary has barely scratched the surface of the extensive literature on contact transformations. The references include more sources of information on these techniques, as does any standard textbook on theoretical or classical mechanics or celestial mechanics.

13. SIMILARITY METHODS

By similarity methods we mean transform methods which combine variables to effect a reduction in order or in the number of variables. These methods sometimes employ separation of variables or physical dimensional analysis, but are primarily concerned with transformation groups. The principle tool is the theory of Lie groups developed by Sophus Lie during the late nineteenth century. We can at best give a little of the flavor of the methods involved in this short section and will attempt no development of Lie theory.

The reader should consult Ames [7, 9], Bluman and Cole [43], Eisenhart [92], Goldstein [122] and their bibliographies for further references.

For first-order ODEs the theory results in solution by quadrature or direct integration after transformation. For higher order ODEs or systems a single transformation attempts to reduce the order or number of equations by one. Successive transformations may eventually reduce the problem to solution by quadrature. For first-order PDEs, characteristic theory reduces the problem to a system of ODEs. However in some cases, similarity methods may produce more easily solved ODEs. See example 9.1. For higher order PDEs or systems, the goal is a reduction in either the number of independent variables or the number of equations (dependent variables).

For simplicity, let us first consider a simple ODE from Bluman and Cole [43, pp. 6–13],

$$y'(x) = F(x, y). \tag{13.1}$$

Let $\beta: (0, \infty) \to (0, \infty)$ be such that

$$\beta(\alpha\gamma) = \beta(\alpha)\beta(\gamma)$$

or equivalently

$$\beta(\alpha) = \alpha^k$$

for some constant k, and consider the transformation

$$T_\alpha(x, y) = (\alpha x, \beta(\alpha)y) = (u, v). \tag{13.2}$$

Note that the restriction on the β means that the mapping $\alpha \mapsto T_\alpha$ from the multiplicative positive reals into the transformations of the plane under composition is a group homomorphism. Thus T_α is called a transformation group. Now the transformation (13.2) changes (13.1) into

$$v'(u) = \beta F(u/\alpha, v/\beta)/\alpha.$$

If this equation has the same form as (13.1), that is,

$$\beta F(u/\alpha, v/\beta) = \alpha F(u, v),$$

then (13.1) is called invariant under T_α. In this case, the above equation may be rewritten as

$$\beta F(x, y) = \alpha F(\alpha x, \beta y).$$

Differentiating with respect to α yields the quasilinear PDE for F,

$$[\beta'(\alpha)\alpha/\beta(\alpha) - 1]F(u, v) = uF_u + [\beta'(\alpha)\alpha/\beta(\alpha)]vF_v.$$

Characteristic theory yields the general solution

$$F(x, y) = x^{(\beta'\alpha/\beta - 1)}G(yx^{-\beta'\alpha/\beta}).$$

Noting that (13.1) does not depend upon α, or considering the form of β, we see that (13.1) must have the form

$$y'(x) = x^{k-1}G(yx^{-k}). \tag{13.3}$$

That is, equations of the special form (13.3) are invariant under the transformation group (13.2). Using this fact a solution is readily obtainable.

Toward this end, let $H = x^k G/y$, and let $\sigma = yx^{-k}$. Then (13.3) becomes

$$y'(x)/y = H(\sigma)/x,$$

or

$$\sigma'(x) = \sigma(H(\sigma) - k)/x,$$

or

$$x'(\sigma)/x = 1/\sigma(H(\sigma) - k) = J(\sigma).$$

This is the desired reduction to quadrature, with solution given by

$$\ln(x/x_0) = \int_{\sigma_0}^{\sigma} J.$$

Let us now turn to similarity for PDEs following Ames [7, pp. 136–144]. For $x \in \mathbb{R}^n$, and $u \in \mathbb{R}^m$, consider the system of m equations

$$F(x, u, u_x) = 0,$$

where $F: \mathbb{R}^{n+m+nm} \to \mathbb{R}^m$, and the one-parameter transformation group

$$\begin{aligned} y^j &= a^{\alpha_j}x^j, & 1 \le j \le n, \\ v^j &= a^{\gamma_j}u^j, & 1 \le j \le m, \end{aligned} \tag{13.4}$$

where the α_j and γ_j are fixed, but will be determined later, and a is the parameter. The α_j and γ_j will be chosen so that F is constant conformally invariant under (13.4). That is,

$$F^j(y, v, v_y) = f_j(a)F^j(x, u, u_x), \qquad 1 \le j \le m. \tag{13.5}$$

This requirement leads to a set of equations which generate "similarity variables," like σ above, which are called the invariants of (13.4).

If x^1 is the variable to be eliminated, there are two cases: either $\alpha_1 = 0$ or $\alpha_1 \ne 0$. In the case $\alpha_1 \ne 0$, Ames lists the invariants as

$$\begin{aligned} \eta_j &= x^j(x^1)^{-\beta_j}, & \beta_j &= \alpha_j/\alpha_1, & 2 \le j \le n, \\ f_j(a) &= u^j(x)(x^1)^{-\gamma_j/\alpha_1}, & & & 1 \le j \le m. \end{aligned} \tag{13.6}$$

If $\alpha_1 = 0$, they are, after redefining the y's and v's,

$$y^1 = x^1 + \ln a,$$
$$y^j = a^{\alpha_j}x^j,$$
$$v^j = a^{\gamma_j}u^j,$$
$$\eta_j = x^j/\exp(\alpha_j x^1), \qquad 2 \le j \le n,$$
$$f_j(a) = u^j(x)/\exp(\gamma_j x^1), \qquad 1 \le j \le m.$$

Example 13.1   Consider the boundary value problem

$$u_t + uu_x = 0,$$
$$u(t, 0) = 0.$$

Let $T = a^p t$, $X = a^q x$, and $U = a^r u$. The preceding PDE becomes

$$a^{p-r}U_T + a^{q-2r}UU_X = 0.$$

If this is $f(a)\,[U_T + UU_X]$, then $f(a) = a^k$ for some k and

$$p - r = k = q - 2r$$

or

$$r = p - k,$$
$$q = 2p - k.$$

To eliminate t, if $b = k/p$, let

$$\eta = xt^{-(2-b)},$$
$$V = ut^{-(1-b)},$$

as in (13.6). These quantities, substituted into the original problem, yield

$$V'(\eta) = \frac{(b-1)V}{(2-b)\eta + V},$$
$$V(0) = 0.$$

For $b = 1$, $V = 0$ and the resulting u is the trivial solution

$$u(t, x) = 0.$$

For $b = 2$, $V' = 1$ so $V = \eta$ and the original problem has nontrivial solution

$$u(t, x) = x/t.$$

For other b, inverting $V(\eta)$ leads to

$$\eta'(V) + [(b - 2)/(b - 1)V]\eta = 1/(b - 1),$$

$$\eta(0) = 0.$$

This simple linear first-order ODE has integrating factor V^μ, where $\mu = (b - 2)/(b - 1)$, and general solution

$$\eta = V/(b - 1)(\mu + 1) + cV^{-\mu}$$

See Spiegel [245, pp. 41–45]. Now $\eta(0) = 0$ is only possible if $\mu \le 0$ or $1 < b < 2$. In this case, u may be determined from

$$(2b - 3)xt^{2b - 4} = ut^{2b - 3} + (2b - 3)cu^{(2 - b)/(b - 1)}$$

in theory. However the initial condition $u(t, 0) = 0$ or $\eta(0) = 0$, puts no restriction on c, and trying simply $c = 0$ does not lead to a solution of the original problem. ∎

We will conclude this section by mentioning briefly the "infinitesimal contact transformations" using the heuristic approach of Goldstein [122, pp. 258–263]. Suppose then that for a generating function of a canonical transformation, we use

$$F(x, q) = xq + \epsilon J(x, q),$$

where J is an arbitrary C^1 function, and ϵ is a "small" parameter. Since the function xq generates the identity transformation, it seems likely that F generates a transformation which varies but slightly from the identity. In fact the transformation is given by

$$p = q + \epsilon J_x(x, q),$$

$$y = x + \epsilon J_q(x, q),$$

$$G = H.$$

As long as J_x and J_q are bounded, it is clear that sufficiently small ϵ makes (y, q) close to (p, x). Assuming that J is reasonably well behaved, so that (y, q) may be uniquely determined for all small ϵ, define

$$x(\epsilon) = y, \qquad x(0) = x,$$

$$p(\epsilon) = q, \qquad p(0) = p.$$

Then

$$\epsilon^{-1}(p(\epsilon) - p(0)) = -J_x(x, p(\epsilon))$$

So, as $\epsilon \to 0$, we see that

$$p'(\epsilon) = -J_x(x, p).$$

Similarly

$$x'(\epsilon) = J_p(x, p).$$

Heuristically, we write

$$dp = -J_x \, d\epsilon,$$

$$dx = J_p \, d\epsilon,$$

and think of $(x, p) \mapsto (x(\epsilon), p(\epsilon))$ as a continuous succession of infinitesimal contact transformation. In particular, if H is independent of t, $J = H$, and ϵ represents time t, we have

$$p'(t) = -H_x(t, p),$$

$$x'(t) = H_p(t, p).$$

These are, of course, just the original Hamiltonian systems, but the solution (x, p) is now thought of as developing in time by a continuous evolution of contact transformations, and the mapping $(x(0), p(0)) \mapsto (x(t), p(t))$ is plausibly expected to be a contact transformation. In fact H is, in a literal sense, the generator of the system evolution in time.

Conversely, the inverse canonical transformation taking $(x(t), p(t))$ back to their constant initial values yields a transformation to constant coordinates $(y, q) = (x(0), p(0))$. Finding such a transformation is clearly equivalent to solving the Hamiltonian system. In fact, we have already seen that a complete integral of the Hamilton–Jacobi equation does indeed act as generating function for such a contact transformation. The above sketch merely indicates that this theory might be developed via infinitesimal contact transformations.

This fascinating area of infinitesimal generators of transformation groups, Lie theory applied to differential equations, similarity transformations, and the study of "orbits," is well worth further investigation by the interested reader. For this reason, many textual references in this area are included at the end of this monograph, and the author genuinely regrets the lack of sufficient space and time to delve further into these subjects here.

II Existence

1. GLOBAL SOLUTIONS

In many physical problems a solution is required for all positive times and at all points of space or at least in a given domain. But the study of characteristics shows that solutions are possible only in a local sense, and that often the domain of existence is very restricted. How can a "global" solution even be considered if that is the case?

Apparently the definition of solution must be relaxed. Some sort of "almost everywhere" solution is a natural requirement. Absolute continuity is a minimum goal, but it turns out that a locally Lipschitzian solution which satisfies the Hamilton–Jacobi equation almost everywhere is usually possible. Thus a global solution is defined to be a locally Lipschitzian real function $u(t, x)$ which satisfies the Hamilton–Jacobi equation at almost every point of the domain in question.

It also turns out that, for very general boundaries, the assumption of boundary values may present a few minor difficulties. Thus a slightly relaxed condition will be required in this area too. To motivate this requirement, consider the following example.

Example 1.1 For the problem in n dimensions,

$$u_t + |u_x|^2 = 0,$$

$$u(0, x) = 0,$$

a solution is given by $u = 0$. Now suppose one additional boundary point
$(1, 0)$ is added. If it is required that

$$u(1, 0) = 0,$$

$u = 0$ is still a solution and no problem develops. But suppose that it is
required that

$$u(1, 0) = -1.$$

Now a solution is given by

$$u(t, x) = \begin{cases} -1 + |x|^2/4(t - 1), & |x|^2 < 4(t - 1), \\ 0, & \text{otherwise.} \end{cases}$$

Here we have a discontinuity in u at $(1, 0)$. ∎

Example 1.1 points out the difficulty in requiring u to be continuous on
the boundary. Note however, that if $(t_k, x_k) \to (1, 0)$ as $k \to \infty$, $u(t_k, x_k) \to -1$
in some cases. In particular, if $|x_k| \le 4(t_k - 1)$ for all large k. For other
approaches, such as $(t_k, x_k) = (1 - 1/k, 0)$, $u(t_k, x_k) \to 0$. Thus we are led to
consider certain types of boundary approaches, called deterministic
approaches, and require only that the boundary data be assumed along
deterministic approaches. For most problems of interest, all or almost all
approaches will be deterministic.

PROBLEM DEFINITION   Let a Hamilton–Jacobi equation

$$u_t + H(t, x, u_x) = 0,$$

be given, along with boundary data, $f \in C(B)$, where B is a closed set in
$\mathbb{R} \times \mathbb{R}^n$, with the infimum of the time components of points of B equal to b.
Let D be the complement of B in $(b, \infty) \times \mathbb{R}^n$. A sequence (t_k, x_k) of points
of D with a limit (t, x) in B will be called a boundary approach and will be
denoted simply by

$$(t_k, x_k) \to (t, x).$$

The boundary approach $(t_k, x_k) \to (t, x)$ will be called *deterministic* if, for
each k, there is a boundary point (r_k, w_k) with $r_k < t_k$, such that, as $k \to \infty$,
$\lim(r_k, w_k) = (t, x)$ and $|x_k - w_k|/(t_k - r_k)$ remains bounded. A solution
of the problem (H, f) is a locally Lipschitzian function $u : D \to \mathbb{R}$ such that
the Hamilton–Jacobi equation is satisfied by u at almost every point of D
and such that, for any deterministic boundary approach $(t_k, x_k) \to (t, x)$,
we have $u(t_k, x_k) \to f(t, x)$ as $k \to \infty$. Here, of course, it is assumed that H is
defined on all of $D \times \mathbb{R}^n$.

Since deterministic boundary approaches are the ones that work, it is desirable to have some readily apparent criteria to tell which approaches are deterministic. Toward this end, nice boundary points are defined. The author has previously used the term "regular boundary point" in this context but cannot bear to pick on the overworked word "regular" further.

DEFINITION   A boundary point (t, x) is called *nice* if it is a limit of a sequence (s_k, y_k) of other boundary points with $s_k < t$, such that as $k \to \infty$, $|x - y_k|/(t - s_k)$ remains bounded.

Alternately stated, a boundary point (t, x) is nice if it is a limit point of the intersection of B with a cone

$$C = \{(s, y): s < t, |x - y|/(t - s) < M\},$$

for some $M > 0$. See figure 1.

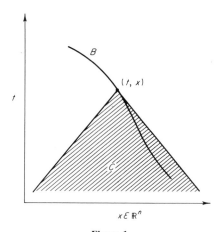

Figure 1

THEOREM 1.1   *Deterministic Boundary Approaches*   Either of the following conditions suffices to ensure that a boundary approach $(t_k, x_k) \to (t, x)$ be deterministic:

(a) (t, x) is a nice boundary point.
(b)   For each $k, t_k > t$ and there is a w_k with $(t, w_k) \in B$, such that as $k \to \infty$, $|x_k - w_k|/(t_k - t)$ remains bounded.

Note that (b) includes the case that $t_k > t$ and either $(t, x_k) \in B, |x_k - x|/(t_k - t)$ remains bounded or (t, x) is in the interior, in $\{t\} \times \mathbb{R}^n$, of $B_t = B \cap \{t\} \times \mathbb{R}^n$. Thus if $B = \{b\} \times \mathbb{R}^n$ (Cauchy problem), every approach is deterministic and if B is a cylinder $\mathbb{R} \times X, X \subset \mathbb{R}^n$, every approach with $t > b$ is deterministic.

Proof (a)   Let $(s_k, y_k) \to (t, x)$ with (s_k, y_k) in B,

$$s_1 < s_2 < \cdots < t,$$

such that, for each k,

$$|x - y_k|/(t - s_k) < M,$$

where $M > 0$ is independent of k. Set

$$m_k = \min\{t_k, 2t_k - t, t - |x - x_k|\}.$$

Note that $m_k \to t$ as $k \to \infty$ and assume, with no loss of generality, that $m_1 > s_1$. For each k, let r_k be the largest s_j such that $s_j \leq m_k$, and let w_k be the corresponding y_j. (If $m_k = t$, let r_k be any $s_j > r_{k-1}$.) Then as $k \to \infty$, $(r_k, w_k) \to (t, x)$. Now note that

$$r_k \leq m_k \leq 2t_k - t,$$

so that

$$t_k - r_k \geq (t - r_k)/2$$

and

$$r_k \leq t - |x - x_k|,$$

so that

$$|x - x_k|/(t - r_k) \leq 1.$$

Thus

$$|x_k - w_k|/(t - r_k) \leq (|x_k - x| + |x - w_k|)/(t_k - r_k)$$
$$\leq 2(|x_k - x| + |x - w_k|)/(t - r_k)$$
$$\leq 2(1 + M).$$

Therefore, $|x_k - w_k|/(t_k - r_k)$ is bounded and the approach is deterministic.

(b)   Simply let $r_k = t$, then (r_k, w_k) works as in the definition of a deterministic approach. ∎

Example 1.2   Consider the problem $u_t + |u_x|^2 = 0$, with $f = 0$ on the boundary, the half sphere

$$B = \{(t, x) \in \mathbb{R} \times \mathbb{R}^n : t \geq 0, \qquad t^2 + |x|^2 = 1\}.$$

This problem has solution

$$u(t, x) = \begin{cases} 0, & |x| \leq 1, \quad t > (1 - x^2)^{1/2}, \\ 1/4t, & x = 0 \quad \text{and} \quad t < 1, \\ |x - x/\|x\||^2/4t, & \text{otherwise.} \end{cases}$$

Note that u is locally Lipschitzian on D and differentiable except for $0 < t < 1, x = 0$. Also f and H are infinitely differentiable. But if $(t_k, x_k) \to (t, x)$ is the boundary approach $(1 - 1/k, 0) \to (1, 0)$, then $u(t_k, x_k) = k/4(k - 1) \to \frac{1}{4}$ as $k \to \infty$, while $f(1, 0) = 0$. So this nondeterministic approach does not assume the boundary data. However, $u(1 + 1/k, 0) = 0 \to f(1, 0)$, so u can not be extended to B in a continuous manner. Of course $(1, 0)$ is not a nice boundary point, nor is $(0, x)$ for any x with $|x| = 1$. The approach $(t_k, x_k) = (1/k^3, (1 + 1/k)x) \to (0, x)$ is not deterministic, and $u(t_k, x_k) \to \infty$ as $k \to \infty$. However all other boundary points are nice. The observant reader may have noticed that $u = 0$ is another solution to this problem, foretelling uniqueness problems to be encountered in the next chapter. ∎

For the remainder of this chapter, the Hamiltonian H will be assumed convex and will always be associated with the Lagrangian L, defined to be the convex dual of H:

$$L(t, x, q) = H^*(t, x, q) = \sup_p\{pq - H(t, x, p)\},$$

and the terms "curve" or "path" will refer to a Lipschitzian function $\alpha : [s, t] \to \mathbb{R}^n$. A curve α will be said to "join" its space-time endpoints $(s, \alpha(s)), (t, \alpha(t))$. It will further be assumed that L is real-valued (finite), thus convex and continuous in q. This is the case if and only if for each (t, x)

$$H(t, x, p)/|p| \to \infty \qquad \text{as} \quad |p| \to \infty.$$

2. THE COMPATIBILITY CONDITION

Now that global solutions have been defined, it is natural to inquire as to what is needed for such a solution to exist. A first necessary condition becomes apparent immediately upon integration between boundary points. If $L = H^*$, the *compatibility condition* will be said to hold if for any (Lipschitzian) path $\alpha : [s, t] \to \mathbb{R}^n$ joining boundary points (t, x) and (s, y),

$$f(t, x) - f(s, y) \leq \int_\alpha L,$$

where

$$\int_\alpha L = \int_s^t L(\tau, \alpha(\tau), \alpha'(\tau)) \, d\tau.$$

If u solves the problem (H, f), then the following sequence of relations implies the compatibility condition:

$$f(t, x) - f(s, y) = u(t, x) - u(s, y), \tag{2.1}$$

$$u(t, x) - u(x, y) = \int_s^t D_\tau u(\tau, \alpha(\tau)) \, d\tau, \tag{2.2}$$

$$D_\tau u(\tau, \alpha(\tau)) = u_t(\tau, \alpha) + u_x(\tau, \alpha)\alpha', \tag{2.3}$$

$$u_t(\tau, \alpha) = -H(\tau, \alpha, u_x(\tau, \alpha)), \tag{2.4}$$

$$u_x(\tau, \alpha)\alpha' - H(\tau, \alpha, u_x(\tau, \alpha)) \le L(\tau, \alpha, \alpha') \tag{2.5}$$

Equations (2.1)–(2.5) yield an englightening heuristic argument to establish the compatibility condition as long as L is measurable, so that L may be integrated along α although the integral may be infinite. However this heuristic procedure contains many gaps. Equation (2.1) is not exactly true since u is not defined on B; (2.2) is only guaranteed if $U(\tau) = u(\tau, \alpha(\tau))$ is absolutely continuous on $[s, t]$; (2.3) only holds in general where u is differentiable; (2.4) holds almost everywhere in D since u solves (H, f), but may hold nowhere along α. Fortunately (2.5) is definition. These gaps may be filled in for reasonably well behaved boundaries, including those of the classical boundary value problem and mixed problems. See Benton [32]. However, this question leads too far from the aim of this chapter to allow further investigation here.

The compatibility condition may be considered a compatibility condition between f and H. Or, for given H, it may be considered a global compatibility condition for f between the different parts of B. On parts of B which are vertical, that is, contain line segments on which x is constant but t varies, and as long as f and B are smooth, the compatibility condition may be interpreted locally as requiring that $-f_t$ is in the range of $H(t, x, \cdot)$, so that if $u(t, x) = f(t, x)$, it is possible to have $u_t + H(t, x, u_x) = 0$. Then if B consists of smooth pieces on which f is also smooth, the compatibility condition becomes a requirement for the piecing together of f on the various parts of B, as well as the local requirement just mentioned on each smooth piece. The next few examples should lead to a better understanding of the compatibility condition.

REMARK *Compatibility for the Cauchy Problem* Let B be contained in the hyperplane $\{b\} \times \mathbb{R}^n$. Then the compatibility condition is vacuously satisfied. This property makes the Cauchy problem much simpler to treat than the general case, as will be noted throughout the study of the Hamilton–Jacobi equation.

Example 2.1 *Compatibility for Plane Boundary* Let e be a unit vector in $\mathbb{R}^n$, Q the hyperplane consisting of vectors x such that $xe = d$, where d is a real constant. Let P be any closed convex subset of Q with relative interior. Let $B = [b, \infty) \times P$. Suppose that f is absolutely continuous and that $H(t, x, p)$ is independent of t and x and convex in p. Then the compatibility condition is implied by the following condition:

$$f_t(t, x) \leq -\min\{H(f_x(t, x) + re) : r \in \mathbb{R}\}$$

almost everywhere on any line segment in B.

To see this, assume this condition and let (t, x) and (s, y) be boundary points with $s < t$. Let $a = (x - y)/(t - s)$ and let λ be the line segment joining (t, x) and (s, y). Then if α is any curve joining (t, x) and (s, y), Jensen's inequality says that

$$(t - s)^{-1} \int_\alpha L \geq L\left((t - s)^{-1} \int_s^t \alpha'\right)$$

$$= L(a).$$

Alternately stated,

$$\int_\alpha L \geq (t - s)L(a)$$

$$= \int_\lambda L.$$

Now letting $g(\tau) = f(\tau, \lambda(\tau))$, g is absolutely continuous since f is absolutely continuous and λ is linear. Therefore for almost all τ,

$$g'(\tau) = f_t(\tau, \lambda(\tau)) + af_x(\tau, \lambda(\tau))$$

$$\leq af_x(\tau, \lambda) + \max\{-H(f_x(\tau, \lambda) + re)\}$$

$$= \max\{af_x - H(f_x + re)\}$$

$$= \max\{a(f_x + re) - H(f_x + re)\}$$

$$\leq L(a)$$

since $a(re) = 0$. This implies that

$$g(t) - g(s) = f(t, x) - f(s, y)$$

$$\leq \int_\lambda L$$

$$\leq \int_\alpha L. \quad \blacksquare$$

Example 2.2   *Compatibility for a Mixed Problem*   With the situation of example 2.1, assume also that inf $L = l > -\infty$ and that there is a constant M such that for x in P and $t > b$,

$$f(t, x) - (t - b)l \leq M.$$

Let G be closed in $\mathbb{R}^n$ and $g \in C(G)$ be bounded below by M such that for $x \in P \cap G$, $f(b, x) = g(x)$. Extend the boundary to include $\{b\} \times G$ and extend f by defining $f(b, x) = g(x)$ for x in G. Then f is continuous and the compatibility condition holds.

To demonstrate this, note that for (t, x), $(s, y) \in B$, $t > s$, the compatibility was shown to hold in example 2.1. Now for x in P and y in G, $t > b$,

$$f(b, y) = g(y) \geq M \geq f(t, x) - (t - b)l,$$

so that

$$f(t, x) - f(b, y) \leq (t - b)l$$

$$\leq (t - b)L((x - y)/(t - b))$$

$$\leq \int_\alpha L$$

for any absolutely continuous curve joining (b, y) and (t, x) again by Jensen's inequality. $\blacksquare$

Example 2.3   *Compatibility for Space Independent Data*   Again let H be convex and independent of t and x, with inf $L = l > -\infty$. Let B be any boundary, with f continuous of the form

$$f(t, x) = \begin{cases} g(t), & t > b, \\ h(x), & t = b, \end{cases}$$

where $g \in C(\mathbb{R})$ with $g(t) - g(s) \leq (t - s)l$ for $b \leq s < t$, and if $b > -\infty$, h is continuous on B_b, and $h \geq g(b)$. Then the compatibility condition holds.

To establish this fact, consider first (s, y), $(t, x) \in B$ with $b < s < t$. Then

$$f(t, x) - f(s, y) = g(t) - g(s)$$
$$\leq (t - s)L((x - y)/(t - s))$$

and the compatibility condition holds for this case again using Jensen's inequality. Now if $s = b$,

$$f(t, x) - f(s, y) = g(t) - h(y)$$
$$\leq g(t) - g(b)$$
$$\leq (t - s)L((x - y)/(t - s)). \qquad ∎$$

3. THE VARIATIONAL SOLUTION

A solution of the problem (H, f) may now be sought. The approach used is an extension of the classical variational approach. To use this approach, assume that L is a measurable function defined on $(b, \infty) \times \mathbb{R}^n \times \mathbb{R}^n$. For any curve $\alpha : [s, t] \to \mathbb{R}^n$ with $y = \alpha(s)$, $(s, y) \in B$, define

$$J(\alpha) = \int_s^t L(\tau, \alpha(\tau), \alpha'(\tau)) \, d\tau. \tag{3.1}$$

To ensure that J is well defined for all α, require also that $L(t, x, p)$ be bounded below by a continuous function l of t. Then $-\infty < J(\alpha) \leq \infty$. Define

$$v(t, x, s, y) = \inf_\alpha J(\alpha). \tag{3.2}$$

Any α such that $v(t, x, s, y) = J(\alpha)$ will be called an extremal, although at this point no claim is made for the existence of such extremals. Finally, for (t, x) a point of B or D, define $u(t, x)$ to be the infimum of $v(t, x, s, y) + f(s, y)$ over all boundary points (s, y) with $s < t$. The following theorem is the critical step in the variational approach and is essentially a case of Bellman's optimality principle. The proof given here is a modification of Fleming's proof given in [109].

THEOREM 3.1 *Hamilton–Jacobi Equation for Variational Minimum* Suppose that $(t, x) \in D$ has an extremal α in any class of curves which includes at least the piecewise linear curves and that u is differentiable at (t, x). Then under the assumption that H and L are upper semicontinuous in (t, x), u satisfies the Hamilton–Jacobi equation at (t, x).

Proof   Let β be any admissible curve joining (t, x) to B. Then for small $\epsilon > 0$,

$$u(t - \epsilon, \beta(t - \epsilon)) + \int_{t-\epsilon}^{t} L(\tau, \beta, \beta') \, d\tau \geq u(t, x).$$

But

$$u(t - \epsilon, \beta(t - \epsilon)) = u(t, x) - \epsilon u_t(t, x) - u_x(t, x) \int_{t-\epsilon}^{t} \beta' + o(\epsilon)$$

by differentiability. Therefore

$$u_t(t, x) + \epsilon^{-1} \int_{t-\epsilon}^{t} \{u_x(t, x)\beta'(\tau) - L(\tau, \beta(\tau), \beta'(\tau)\} \, d\tau \leq o(1).$$

Now let $q \in \mathbb{R}^n$ be arbitrary and choose β with $\beta' = q$ near t. Then letting $\epsilon \to 0$, one has

$$u_t(t, x) + u_x(t, x)q - \lim \inf \epsilon^{-1} \int_{t-\epsilon}^{t} L(\tau, \beta(\tau), q) \, d\tau \leq 0.$$

This implies that

$$\lim_{\tau \to t} \inf L(\tau, \beta(\tau), q) \geq u_t(t, x) + u_x(t, x)q.$$

Since L is upper semicontinuous in (t, x),

$$L(t, x, q) \geq u_t(t, x) + u_x(t, x)q,$$

or

$$u_x(t, x)q - L(t, x, q) \leq -u_t(t, x).$$

Since this holds for all q in $\mathbb{R}^n$,

$$H(t, x, u_x(t, x)) \leq -u_t(t, x).$$

On the other hand, letting $\beta = \alpha$, one sees that

$$u_t(t, x) + \epsilon^{-1} \int_{t-\epsilon}^{t} \{u_x(t, x)\alpha'(\tau) - L(\tau, \alpha(\tau), \alpha'(\tau))\} \, d\tau = o(1).$$

As a consequence,

$$\lim_{\tau \to t} \sup\{u_x(t, x)\alpha'(\tau) - L(\tau, \alpha(\tau), \alpha'(\tau))\} = M \geq -u_t(t, x).$$

Thus by the definition of L^*,

$$\lim_{\tau \to t} \sup H(\tau, \alpha(\tau), u_x(t, x)) \geq M.$$

Since H is upper semicontinuous in (t, x).

$$H(t, x, u_x(t, x)) \geq M \geq -u_t(t, x).$$

The opposite inequality has already been obtained, so that

$$u_t(t, x) + H(t, x, u_x(t, x)) = 0$$

as was to be shown. ∎

This theorem indicates that if L and H are upper semicontinuous in (t, x) and measurable and if u is differentiable and has an extremal at almost every point (t, x) in D, then u is an almost everywhere solution of the Hamilton–Jacobi equation in D. (Note that the upper semicontinuity of H implies the lower semicontinuity of L in t and x since it is a supremum of lower semicontinuous functions. Thus we are really requiring the continuity of both L and H in t and x. For the remainder of this chapter it will be assumed that L is continuous in all variables and that H is at least upper semicontinuous in all variables.) If u is locally Lipschitzian and if the boundary approaches behave correctly, it follows that u solves the boundary value problem (H, f). The next few sections will display conditions which guarantee that this is indeed the case.

4. GROWTH CONDITIONS

To ensure the existence of a variational minimum, and to make the boundary approaches behave correctly, it is convenient to introduce a growth condition at this point. This is simply that $L(t, x, q)/|q| \to \infty$ as $|q| \to \infty$ uniformly in t and x, and will be assumed to hold throughout the remainder of this chapter. Requiring that $H(t, x, p)$ be bounded above for bounded p uniformly in (t, x) is sufficient to guarantee the growth condition on L. To see this, note that for any p in $\mathbb{R}^n$,

$$L(t, x, q) \geq qp - H(t, x, p).$$

If $M > 0$ is arbitrary, letting $p = Mq/|q|$,

$$L(t, x, q)/|q| \geq M - H(t, x, Mq/|q|)/|q|$$

$$\geq M - \sup\{H(T, X, P) : |P| = M\}/|q|.$$

Thus as $|q| \to \infty$, $\liminf L(t, x, q)/|q| \geq M$. Since M was arbitrary, this implies the growth condition for L. A slight variation of this demonstration establishes the earlier contention that L is finite if and only if $H(t, x, p)/|p| \to \infty$ as $|p| \to \infty$ for each fixed (t, x).

Recalling that L is assumed to be bounded below by the continuous function l of t, we have the following key consequence of the growth condition.

LEMMA 4.1   For each k, let α_k be a curve joining (s_k, y_k) to (t_k, x_k) with $s_k < t_k$. Assume either that t_k and s_k remain bounded or that l may be chosen to be ≥ 0. Then if $|x_k - y_k|$ is bounded away from zero but $a_k = |x_k - y_k|/(t_k - s_k) \to \infty$ as $k \to \infty$, it follows that

$$\int_{\alpha_k} L \to \infty.$$

Proof   Assume that for all k, $|x_k - y_k| > 2\delta > 0$. Now let $M > 0$ be given. Suppose that for $|q| > N$, $L(t, x, q)/|q| > M$. Choose a K such that for $k > K$, $a_k > 2N$. Now fix $k > K$ and let $S = (s_k, t_k) \subset \mathbb{R}$. Let P be the subset of S consisting of those t with $|\alpha_k'(t)| < a_k/2$. Define Q to be the complement of P in S. Then

$$\int_P |\alpha_k'| < a_k(t_k - s_k)/2,$$

so that

$$\int_Q |\alpha_k'| = \int_S |\alpha_k'| - \int_P |\alpha_k'|$$

$$> a_k(t_k - s_k)/2.$$

Certainly l may be chosen to be ≤ 0, so that

$$\int_{\alpha_k} L \geq \int_P l + \int_Q L$$

$$\geq \int_S l + M\delta.$$

Now letting $k \to \infty$, $\int_S l$ remains bounded since by hypothesis either (s_k, t_k) remains bounded or $l = 0$. Since M was arbitrary, $\int_{\alpha_k} L \to \infty$ as claimed.   ∎

The next two lemmas establish the boundary approach properties. For both it will be assumed that $(t_k, x_k) \to (t, x)$ is a deterministic boundary approach with $(r_k, w_k) \to (t, x)$ from B, $r_k < t_k$, and

$$a_k = |x_k - w_k|/(t_k - r_k)$$

$$< m,$$

for some fixed $m > 0$. It is also assumed that for each k, (t_k, x_k) has a foot point (s_k, y_k), that is,

$$u(t_k, x_k) = f(s_k, y_k) + v(t_k, x_k, s_k, y_k).$$

LEMMA 4.2   If $(s_k, y_k) \to (t, x)$ as $k \to \infty$, then $u(t_k, x_k) \to f(t, x)$. In any case,

$$\limsup_{k \to \infty} u(t_k, x_k) \le f(t, x).$$

Proof   Let λ_k be the line segment joining (r_k, w_k) to (t_k, x_k). Then

$$u(t_k, x_k) - f(r_k, w_k) \le v(t_k, x_k, r_k, w_k)$$
$$\le J(\lambda_k).$$

But along the curves λ_k, all arguments of L are bounded, so that L is bounded and $J(\lambda_k) \to 0$ as $k \to \infty$. This implies the last assertion of the theorem. But if $(s_k, y_k) \to (t, x)$, then

$$f(s_k, y_k) + \int_{s_k}^{t_k} l \le u(t_k, x_k),$$

which shows that

$$\lim f(s_k, y_k) = f(t, x) \le \liminf u(t_k, x_k)$$

since $\int l \to 0$. ∎

For the remainder of this chapter we will assume one further mild growth condition, this one on the data. This is, if $b = -\infty$, so that our boundary has arbitrarily large negative times, then

$$f(s, y) + \int_s^0 l \to \infty \qquad \text{as} \quad s \to -\infty$$

uniformly in y. This forces $f(s, y) + v(t, x, s, y)$ to grow without bound as $s \to -\infty$ uniformly in x and y as long as t remains bounded. Thus for minimizing this function in a neighborhood of (t, x), only (s, y) with s bounded need be considered. For simplicity then, in all proofs it will be assumed that $t > s > 0$.

LEMMA 4.3   Suppose that the compatibility and growth conditions hold and that $v(t, x, s, y)$ is continuous. Then as $k \to \infty$,

$$u(t_k, x_k) \to f(t, x).$$

Proof Assume by way of contradiction that $u(t_k, x_k)$ does not approach $f(t, x)$ as $k \to \infty$. Then by lemma 4.2, (s_k, y_k) cannot tend to (t, x). But the sequence (t_k, x_k) is bounded and without loss of generality, $0 \le s_k < t_k$. If $a_k = |x_k - y_k|/(t_k - s_k)$ is unbounded, then $u(t_k, s_k)$ is unbounded by lemma 4.1. Lemma 4.2 does not allow this, so a_k is bounded. Thus the sequence (s_k, y_k) is bounded and has a limit point, which must be in B since B is closed. Choosing a subsequence if necessary, assume that $(s_k, y_k) \to (s, y) \ne (t, x)$. Since a_k is bounded, $a = |x - y|/(t - s)$ is finite, so that $s < t$. By continuity $u(t_k, x_k) \to v(t, x, s, y) + f(s, y)$. By lemma 4.2, this limit is not greater than $f(t, x)$. But equality is excluded since $u(t_k, x_k)$ does not tend to $f(t, x)$. Therefore

$$f(s, y) + \inf_\alpha J(\alpha) < f(t, x),$$

contradicting the compatibility condition. ∎

5. REGULARITY

In this section, a condition which ensures the necessary regularity of u and v will be considered. For want of a better name, this will be referred to as the regularity condition. The regularity condition will be said to hold if for each (t, x) in B or D and each (s, y) in B with $s < t, v(t, x, s, y)$ may be obtained by minimizing only over those curves whose graphs are contained in the intersection of two cones of the form:

$$C = \{(r, w): |x - w| \le \rho(t - r), r \le t\},$$
$$K = \{(r, w): |w - y| \le \rho(r - s), s \le r\},$$

where ρ may be chosen to vary in a locally bounded manner with (t, x, s, y).

LEMMA 5.1 Let $(T, X) \in B \cup D, (S, Y) \in B, S < T$. Then as long as the regularity condition holds, there are neighborhoods V of (T, X) and W of (S, Y) such that $v(t, x, s, y)$ is Lipschitzian in $(t, x) \in V$ uniformly in $(s, y) \in W$ and vice versa.

Proof Choose V and W small enough so that $(t, x) \in V$ and $(s, y) \in W$ implies that $t - s > \epsilon$ for some fixed $\epsilon > 0$. Let the same ρ work as in the above definition for all $(t, x) \in V$ and all $(s, y) \in W$. Now let (s, y) be fixed in W and (t, x) and (r, w) be points of V with $r \le t$. Let d denote the Euclidean distance from (t, x) to (r, w). Let $\{\alpha_k\}$ and $\{\beta_k\}$ be minimizing sequences in the proper cone as guaranteed by the regularity condition. That is, α_k joins (t, x) and (s, y), β_k joins (r, w) and (s, y), and

$$J(\alpha_k) \leq v(t, x, s, y) + 1/k,$$

$$J(\beta_k) \leq v(r, w, s, y) + 1/k.$$

Make V and W smaller if necessary, so that $r - d$ is greater than s and $d < 1$. Let λ^k be the line segment joining $(r - d, \beta_k(r - d))$ and (t, x), and let λ_k be that connecting $(r - d, \alpha_k(r - d))$ to (r, w). See figure 2. Then

$$v(t, x, s, y) \leq J(\beta_k) - \int_{r-d}^{r} L(\tau, \beta_k, \beta_k')\, d\tau + \int_{\lambda_k} L$$

$$\leq v(r, w, s, y) + Md + \int_{\lambda_k} L + 1/k,$$

where M is a bound for $|l|$ for all time components considered.

   Now by the regularity condition, α_k and β_k were chosen so that L is bounded above on the graphs of λ_k and λ^k by a constant depending only

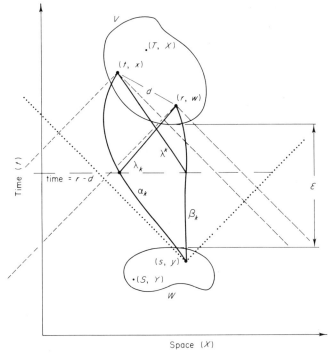

Figure 2 $(-\, -\, -)$ boundary of cones C; $(\ldots)$ boundary of cone K.

upon V and W. Assume with no loss of generality that this constant is also M, increasing M if necessary. Then

$$v(t, x, s, y) \le v(r, w, s, y) + Md + (t - r + d)M + 1/k$$

$$\le v(r, w, s, y) + 3Md + 1/k.$$

Similarly

$$v(r, w, s, y) \le v(t, x, s, y) + 3Md + 1/k.$$

Letting $k \to \infty$, one sees that $3M$ is a Lipschitz constant for $v(t, x, s, y)$ as a function of (t, x) in a neighborhood of (T, X). With no loss of generality, this neighborhood is V. But M does not depend upon $(s, y) \in W$, only upon V and W. Thus $v(t, x, s, y)$ is Lipschitzian in (t, x) uniformly in $(s, y) \in W$ as claimed. For the vice versa claim, a similar proof interchanging the roles of (t, x) and (s, y) works. ∎

Now since f is continuous and the regularity condition yields the continuity of $v(t, x, s, y)$, all that is needed to make the boundary approaches behave correctly is the growth conditions. We are led to the following basic theorem.

THEOREM 5.1   *General Variational Existence Theorem*   Let (H, f) be a Hamilton–Jacobi initial value problem, where $H = L^*$ with $L(t, x, q)$ continuous, convex in q, and bounded below by $l(t)$, where l is continuous for $t \ge b$. Suppose that the compatibility, regularity, and growth conditions all hold. Suppose also that each fixed endpoint variational problem has an extremal in the given class of admissible curves (as previously discussed), and that $f(s, y)$ is bounded below for bounded s uniformly in y. Then

$$u(t, x) = \min\left\{ f(s, y) + \int_s^t L(\tau, \alpha, \alpha') \, d\tau \right\}$$

is a solution of the problem (H, f), where the minimum is taken over all admissible curves joining $(s, y) \in B$, with $s < t$, to (t, x).

Proof   Let (t, x) be a point of D. Choose a minimizing sequence $\{\alpha_k\}$ such that

$$f(s_k, y_k) + J(\alpha_k) \le u(t, x) + 1/k,$$

where α_k joins $(s_k, y_k) \in B$ to (t, x), and such that each α_k is an extremal for its associated fixed endpoint problem. Now if $a_k = |x - y_k|/(t - s_k)$ becomes unbounded, by choosing a subsequence if necessary, it may be assumed that $a_k \to \infty$ as $k \to \infty$. But the distance from (t, x) to the boundary is positive, so that $a_k \to \infty$ implies that $|x - y_k|$ is bounded away from zero. By the second growth condition, we assume as before that $0 \le s_k < t$. Then by the first growth condition and lemma 4.1, it follows that $J(\alpha_k)$ becomes unbounded. This contradicts the above inequality since f is bounded below for bounded s_k by hypothesis. Thus we see that a_k remains bounded as $k \to \infty$. This in turn implies that (s_k, y_k) remains bounded. Thus this sequence has a limit point $(s, y) \in B$ with $s < t$. Letting α be an extremal joining (s, y) to (t, x), we see that $u(t, x) = f(s, y) + J(\alpha)$. Thus each point (t, x) of D has an extremal.

Now let V be a small neighborhood of (t, x) contained in D, with T the infimum of time components of points of V and (S, Y) a fixed boundary point with $S < T$. For all $(r, w) \in V, u(r, w)$ is bounded above by $f(S, Y) + v(r, w, S, Y)$. Let $a(r, w)$ be $|w - y|/(r - s)$, where (s, y) is the foot point of an extremal for (r, w). On V, the function $a(r, w)$ is bounded, for if not, lemma 4.1 contradicts the fact that $u(r, w)$ is bounded above. Thus for (r, w) in V, $u(r, w)$ may be obtained by minimizing $f(s, y) + v(r, w, s, y)$ over some compact set K of points (s, y) of B.

The proof of lemma 5.1 may be repeated with K replacing W for sufficiently small V. See exercise 5.1. This establishes the fact that u is locally Lipschitzian. It might be pointed out that this proof will not work if V becomes too large. In fact, it is generally impossible to obtain a Lipschitzian solution.

Since u is locally Lipschitzian, it is differentiable almost everywhere by Rademacher's theorem. Since each point of D has an extremal, theorem 3.1 applies almost everywhere in D and u is an almost everywhere solution of the Hamilton–Jacobi equation. By the growth and regularity conditions, lemma 4.3 applies and the compatibility condition implies that the boundary approach behavior is correct. Thus u solves the problem (H, f) as was to be shown. ∎

Of course this theorem does not completely solve the question of existence when confronted with a particular problem (H, f). However, it has reduced it to a variational problem. The existence of extremals for fixed endpoint problems with properly bounded derivatives will demonstrate that the regularity condition holds and that u is indeed a minimum rather than just an infimum. Of course, this last condition is needed only on a set of full measure, but it will be just as easy to prove it at all points by placing general conditions on H. The next few sections will solve this variational problem for various cases, thus yielding fairly general existence theorems.

Exercise

5.1   Prove that u is locally Lipschitzian if the growth conditions hold and extremals always exist.

   Hint   As in lemma 5.1, vary the extremals to two points of V by small straight line segments joining each point to the other extremal.

6. SPACE–TIME INDEPENDENT HAMILTONIANS

As example 2.2 indicates, the easiest variational problem to handle is the case where $H(t, x, p)$ is independent of t and x. Then if H is convex, Jensen's inequality shows that straight line segments are extremals, and $v(t, x, s, y)$ may be evaluated without reference to integration. In fact, this case may actually be considered to be a use of the complete integral rather than a variational technique. See Hopf [144] and Aizawa and Kikuchi [3].

Recall that, if H is real, convex, and satisfies the growth condition $H(p)/|p| \to \infty$ as $|p| \to \infty$, then $L = H^*$ has the same properties and both are continuous. By the above growth condition for L it follows that L is bounded below by a constant l. The following theorem provides a solution in this case.

THEOREM 6.1   *First Existence Theorem*   Let $H : \mathbb{R}^n \to \mathbb{R}$ be convex with convex dual $H^* = L$. Let B be a closed set in $\mathbb{R} \times \mathbb{R}^n$ with b the infimum of time components of B and $f \in C(B)$ such that the following conditions hold:

   (H1) $H(p)/|p| \to \infty$ as $|p| \to \infty$.
   (D1) $f(t, x) - f(s, y) \le (t - s)L((x - y)/(t - s))$,
               for $(t, x), (s, y) \in B, s < t$.
   (D2) $f(s, y) \ge g(s), g \in C(\mathbb{R})$.
   (D3)   If $b = -\infty$, then $f(s, y) - (\min L)s \to \infty$ as $s \to -\infty$, uniformly in y.

Then

$$u(t, x) = \min\{f(s, y) + (t - s)L((x - y)/(t - s)) : (s, y) \in B, s < t\}$$

is a solution of the problem (H, f). That is, u is locally Lipschitzian, satisfies the Hamilton–Jacobi equation almost everywhere in $D = (b, \infty) \times \mathbb{R}^n$, and assumes the boundary data f along any deterministic boundary approach. The conclusion still holds if **(D2)** is replaced by the slightly weaker condition:

(**D2′**)   As long as t, x, and s remain bounded,

$$f(s, y) + (t - s)L((x - y)/(t - s)) \to \infty$$

as $|y| \to \infty$ uniformly in t, x, and s.

Proof   Since straight line segments are extremals, the regularity condition holds trivially. Integrating along a straight line segment, one sees that (**D1**) is the compatibility condition and that u defined above is the variational minimum. (**H1**) for L and (**D3**) are the growth conditions and (**D2**) makes $f(s, y)$ properly bounded below. Thus theorem 5.1 applies.

Now let us consider (**D2′**). By the growth condition $L(q)/|q| \to \infty$ as $|q| \to \infty$, the second term in (**D2′**) goes to infinity correctly. Thus if (**D2**) holds, (**D2′**) holds trivially. But examination of the proof of theorem 5.1 reveals that (**D2′**) is really all that is required, rather than (**D2**).   ∎

COROLLARY 6.1   Consider the Cauchy problem

$$u_t(t, x) + H(u_x(t, x)) = 0,$$

$$u(0, x) = f(x), \qquad x \in B \subset \mathbb{R}^n,$$

with f continuous, B closed, H convex, and $H(p)/|p| \to \infty$ as $|p| \to \infty$. If f is bounded below, or at least $f(y) + tL((x - y)/t) \to \infty$ as $|y| \to \infty$ uniformly in (t, x) as long as (t, x) remains bounded, then

$$u(t, x) = \min\{f(y) + tL((x - y)/t) : y \in B\}$$

is a solution of this Cauchy problem.   ∎

Example 6.1   Let $H(p) = |p|^2$, so that $L(q) = |q|^2/4$. Then the above Cauchy problem has a solution as long as f is continuous and $f(y) + |x - y|^2/4t \to \infty$ as $|y| \to \infty$ for each t and x, with $t > 0$. In this case the solution is given by

$$u(t, x) = \min_{y}\{f(y) + |x - y|^2/4t\}.$$

Compare this with the examples of chapter 1, where only linear and quadratic f could be handled easily and global solutions could only be obtained by finding a formula for u and showing that the formula was valid globally. Note that even the data of example 8.5 of chapter I, $f(x) = \sin(x/\epsilon)$, yields a global solution by the variational method, although of course the global solution cannot be smooth.   ∎

COROLLARY 6.2   Let e be a unit vector in $\mathbb{R}^n$, Q the hyperplane $\{x \in \mathbb{R}^n : xe = d\}$, P any closed convex subset of Q. Let B be $[b, \infty) \times P$, f a continuously differentiable function on B which is bounded below by a continuous function of t. If $H : \mathbb{R}^n \to \mathbb{R}$ is convex and $H(p)/|p| \to \infty$ as $|p| \to \infty$, let $L = H^*$. If

$$f_t(t, x) \le -\min\{H(f_x(t, x) + re) : r \in \mathbb{R}\}$$

for all t, x, then

$$u(t, x) = \min\{f(s, y) + (t - s)L((x - y)/(t - s)) : (s, y) \in B, s < t\}$$

solves the problem (H, f) and u may be extended to all of $(b, \infty) \times \mathbb{R}^n$ in a continuous manner.

Proof Theorem 6.1 and example 2.2. For the last claim above, simply note that all points of B are nice except for those in B_b. ∎

COROLLARY 6.3   In corollary 6.2, let B also include $\{b\} \times G$, with G closed in $\mathbb{R}^n$, $g \in C(G)$ bounded below by M, $f(b, x) = g(x)$, and if $l = \min L$, assume that

$$f(t, x) - (t - b)l \le M$$

for $t > b$. Then u solves the mixed problem (H, f).

Proof Theorem 6.1 and example 2.3. ∎

COROLLARY 6.4   *Space Independent Boundary Data*   Let $H : \mathbb{R}^n \to \mathbb{R}$ be convex with $H(p)/|p| \to \infty$ as $|p| \to \infty$, and let $L = H^*$, $\min L = l$. Let B be any closed set in $\mathbb{R} \times \mathbb{R}^n$, $f \in C(B)$ of the form

$$f(t, x) = \begin{cases} g(t), & t > b, \\ h(x), & t = b, \end{cases}$$

where for $b \le s < t$,

$$g(t) - g(s) \le (t - s)l. \tag{6.1}$$

If $b = -\infty$, of course h is empty, but if $b > -\infty$, require also that

$$h(x) \ge g(b), \qquad (b, x) \in B.$$

Then u solves the problem (H, f). Note also that if g is absolutely continuous, (6.1) may be replaced by $g'(t) \le l$ almost everywhere.

Proof Theorem 6.1 and example 2.4. ∎

7. SPACE–TIME DEPENDENCE

Section 6 demonstrated the ease with which the space–time independent case can be handled. It seems likely that a small amount of space–time dependence could still be handled, but perhaps not with as much ease. In this section, general space–time dependence is introduced but is carefully bounded. The problem is still solvable, but with greatly increased difficulty. The technique used will be the so-called direct method of the calculus of variations employing lower semicontinuity and Ascoli's theorem.

Lower semicontinuity is the crucial element of the direct method and follows from convexity quite easily. Thus we begin with the following lemma.

LEMMA 7.1   Let $\alpha_k \to \alpha$ uniformly, where $\alpha, \alpha_k : [s, t] \to \mathbb{R}^n$. Suppose that $L(t, x, q)$ is locally Lipschitzian in x uniformly in (t, q) as long as (t, q) remains bounded, and that the α_k's are all Lipschitzian with the same constant K. Let L be convex in q and let L_q exist everywhere and be continuous. Then $J(\alpha) \leq \lim \inf J(\alpha_k)$.

Proof   Break $J(\alpha)$ up as follows:

$$J(\alpha) = J(\alpha_k) + \int_s^t \{L(\tau, \alpha, \alpha_k') - L(\tau, \alpha_k, \alpha_k')\} \, d\tau$$

$$+ \int_s^t \{L(\tau, \alpha, \alpha') - L(\tau, \alpha, \alpha_k')\} \, d\tau.$$

Assuming with no loss of generality that K is also a Lipschitzian constant for $L(t, x, q)$ in x for the arguments involved, the first integral above is bounded in absolute value by $K(t - s)\|\alpha_k - \alpha\|_\infty$, which tends to zero as k becomes infinite. Thus

$$J(\alpha) \leq \lim \inf J(\alpha_k) + \lim \sup \int_s^t (L - L_k),$$

where L represents $L(\tau, \alpha(\tau), \alpha'(\tau))$ and L_k denotes $L(\tau, \alpha(\tau), \alpha_k'(\tau))$. Now, by convexity, wherever α' and α_k' both exist, which is almost everywhere on $[s, t]$, then

$$L - L_k \leq L_q(\tau, \alpha(\tau), \alpha'(\tau))(\alpha'(\tau) - \alpha_k'(\tau)).$$

But $\alpha_k' \to \alpha'$ weakly and α' is bounded and measurable, so that $L_q(\tau, \alpha(\tau), \alpha'(\tau))$ is integrable. Thus $\lim \sup \int_s^t (L - L_k)$ is bounded above by

$$\lim \int (L_q)(\alpha' - \alpha_k') = 0.$$

To check this statement, approximate L_q by a mollifying function and integrate by parts. Therefore,

$$J(\alpha) \leq \lim \inf J(\alpha_k)$$

as claimed. ∎

There are several forms of this lemma, some requiring lower semi-continuity only at a minimizing sequence, which is actually all that is needed, and some not requiring that α be Lipschitzian. For example, see Akhiezer [5]. However, the preceding lemma is all that we shall require.

Let us now turn our attention to the (t, x) dependence of H and L. Since all paths use t as a parameter, the t dependence has no influence on the extremals. However x dependence is quite another story. If we consider momentarily the problem of driving by automobile from one point to another so as to minimize cost, it is clear that a large pile of free gold a few miles out of our way would have a great effect upon our extremals. In fact it would act rather like a magnet, drawing all nearby extremals to itself. For our slightly more abstract problem, the same phenomenon can occur. That is, a steep gradient in L in its space variables can greatly warp the extremals. The success of our method requires extremals with properly bounded "slope," $|\alpha'|$ or properly limited curvature. Thus we must limit the x dependence to some extent. The precise limitation will appear in the next paragraph.

The Hamiltonian, $H(t, x, p)$, will now be assumed to be strictly convex in p, continuously differentiable, and will be required to satisfy the growth condition $H(t, x, p)/|p| \to \infty$ as $|p| \to \infty$ for each fixed (t, x). Recall that these three properties together imply that $L = H^*(t, x, q)$ is finite, continuously differentiable, strictly convex in q, and that

$$L(t, x, q) = pH_p(t, x, p) - H(t, x, p),$$

where

$$q = H_p(t, x, p),$$

$$p = L_q(t, x, q).$$

Also $H_x(t, x, p) = -L_x(t, x, q)$, where p and q are related as in the foregoing. To insure that L satisfies the required growth condition, we shall require further that $H(t, x, p)$ is bounded above by continuous function of p. The restriction on space dependence is

$$|H_x(t, x, p)| \leq g_1(t)(pH_p - H) + g_2(t),$$

where g_1 and g_2 are continuous, g_1 nonnegative. This yields $|L_x| \leq g_1 L + g_2$. It will also be assumed that $pH_p - H \geq g_3(t)$, g_3 continuous, so that $L \geq g_3 = l$, and that

$$|H_p(t, x, p)| \leq g_4(|p|),$$

where $g_4 > 0$ is increasing on $[0, \infty)$.

Now consider the fixed endpoint problem of minimizing J over the set of Lipschitzian curves joining (t, x) with (s, y) with $s < t$. Temporarily add the restraint that each curve have Lipschitz constant $K \leq r$, where $r \geq |x - y|/(t - s)$ is fixed. Call the infimum for this constrained problem $v_r(t, x, s, y)$.

LEMMA 7.2   The constrained problem has an extremal.

Proof   See exercise 7.1.   ∎

LEMMA 7.3   The unconstrained fixed endpoint problem has an extremal in the class of all Lipschitzian curves.

Proof   For each r, let α_r be an extremal for the corresponding constrained problem. Integrating along α_r, define $h(\tau) = \int_s^\tau L_x$. By Pontryagin's maximum principle, it follows that, for almost all τ, the quantity $qh(\tau) - L(\tau, \alpha_r(\tau), q)$ is maximized on the set of all q with $|q| \leq r$ at $q = \alpha_r'(\tau)$. Now by our assumptions on H,

$$|h(\tau)| \leq \int_s^\tau g_1 L + g_2$$

$$\leq (\max g_1)\left\{\int_s^t (L - l) + \int_s^t |l|\right\} + (\max |g_2|)(t - s).$$

Now this last expression is bounded above by the corresponding expression where the integration is along the line segment joining (t, x) to (s, y). This yields a bound on h which we will denote by $M(t, x, s, y)$, which is a continuous function of (t, x) and (s, y). Therefore,

$$|H_p(\tau, \alpha_r(\tau), h(\tau))| \leq g_4(M(t, x, s, y)).$$

But $q = H_p(\tau, \alpha_r(\tau), h(\tau))$ is the point where $qh - L$ is maximized on all of $\mathbb{R}^n$. Thus if $r > g_4(M)$, $\|\alpha_r'\|_\infty \leq g_4(M)$, which is independent of r, so that $\alpha = \alpha_r$ gives the minimum without the constraint $|\alpha'| \leq r$. That is, α is an extremal in the set of all Lipschitzian curves joining (t, x) to (x, y).   ∎

Putting together these three lemmas and theorem 5.1, we obtain the following quite general existence theorem.

THEOREM 7.1 *Second Existence Theorem* Let (H, f) be a Hamilton–Jacobi initial value problem, with b the infimum of the time components of the boundary B. Let u be defined by

$$u(t, x) = \min\{f(s, y) + J(\alpha): (s, y) \in B, s < t, \alpha \text{ admissible}\},$$

where the admissible curves are the Lipschitzian curves joining (t, x) to (s, y). Then u solves the problem (H, f) if the following conditions all hold:

  (H1) $H(t, x, p)$ is strictly convex in p.
  (H2) H is continuously differentiable.
  (H3) $H(t, x, p)/|p| \to \infty$ as $|p| \to \infty$ for each (t, x).
  (H4) $|H_x| \le g_1(t)(pH_p - H) + g_2(t)$.
  (H5) $pH_p - H \ge g_3(t)$.
  (H6) $H, |H_p| \le g_4(|p|)$.
  (D1) f is continuous.
  (D2) for any Lipschitzian curve α joining $(t, x) \in B$ with $(s, y) \in B$, $s < t$, $f(t, x) - f(s, y) \le \int_\alpha L$.
  (D3) $f(s, y) \ge g_5(s)$.
  (D4) if $b = -\infty$, then $f(s, y) + \int_s^0 g_3 \to \infty$ as $s \to -\infty$ uniformly in y.

Here g_i is a continuous real function, $1 \le i \le 5$, g_1 and g_4 are nonnegative, and g_4 is increasing. (L and J are as previously defined.)

Proof By lemma 7.3, each fixed endpoint problem has a solution and an extremal. In the proof of lemma 7.3, the bound on $\|\alpha'\|_\infty$, $g_4(M)$, is a continuous function of the endpoints. This implies the regularity condition. Thus all of the hypotheses of theorem 5.1 are satisfied. ∎

COROLLARY 7.1 If B is the hyperplane $t = 0$, then the problem (H, f) is solved by u, the variational minimum, if f is continuous and bounded below and **(H1)**–**(H6)** all hold. ∎

Example 7.1 Consider the Hamiltonian

$$H(t, x, p) = g(x)(h(t) + |p|^2)^{v/2},$$

where $v > 1$, g and h are continuously differentiable, with $0 < a \le g, h \le b$, and $|g_x|$ is bounded. It may be shown that **(H1)**–**(H6)** are satisfied for this H, so that theorem 7.1 applies to any data f satisfying **(D1)**–**(D4)**. This example is from Fleming [109], in slightly modified form. To establish the compatibility condition, one may establish the stronger condition

$$f(t, x) - f(s, y) \le -(t - s)b^{(1 + v/2)}$$

for $(t, x), (s, y) \in B, s < t$. ∎

Exercises

7.1 Prove lemma 7.2 by the direct method. That is, choose a minimizing sequence, then show that Ascoli's theorem applies, yielding a sub-sequence which converges to a suspected extremal. Show that the suspected extremal is admissible (Lipschitzian with constant $\leq r$) and apply lower semicontinuity.

7.2 Show that the Hamiltonian in example 7.1 satisfies **(H1)**–**(H6)** of theorem 7.1.

8. EQUIVALENT PROBLEMS

Using theorem 5.1 or similar methods we see that variational existence theorems lead to existence theorems for Hamilton–Jacobi initial value problems. There are now many quite general existence theorems for variational problems, most of them coming from control theory or differential geometry. However, their study would lead us too far astray. Thus we will content ourselves with theorem 7.1. However, one very simple observation allows a slight extension of theorems 6.1 and 7.1. That is, that the addition of an exact integrand does not affect the variational extremals since the integral of the exact integrand is independent of the path used. This fact is stated precisely in the following lemma.

LEMMA 8.1 Let (H, f) be an initial value problem, B the closed domian of f, $L = H^*$, and w a locally Lipschitzian function on $B \cup D$. Then $u: D \to \mathbb{R}$ solves the problem (H, f) if and only if $U = u + w$ solves the problem $(\mathscr{H}, \varphi)$, where $\varphi = f + w$ and

$$\mathscr{H}(t, x, p) = H(t, x, p - w_x(t, x)) - w_t(t, x).$$

Also $\mathscr{L} = \mathscr{H}^*$ is given by

$$\mathscr{L}(t, x, q) = L(t, x, q) + w_t(t, x) + qw_x(t, x).$$

Thus u is the variational solution of (H, f) if and only if U is the variational solution of $(\mathscr{H}, \varphi)$, as long as w is differentiable almost everywhere along each curve $\alpha: [s, t] \to \mathbb{R}^n$.

Proof Exercise 8.1 ▮

This lemma extends the previous existence theorems by allowing one to start with a problem not satisfying the necessary hypotheses and adding an exact integrand to produce an equivalent problem which does satisfy the hypotheses. Of course this is the simple transformation introduced in section 9 of chapter I, except that now w is allowed to be locally Lipschitzian but not necessarily differentiable everywhere.

Example 8.1   Consider the Cauchy problem in n dimensions,

$$u_t(t, x) + |2tx + u_x(t, x)|^2 + |x|^2 = 0,$$

$$u(0, x) = f(x),$$

where f is continuous and bounded below. Letting

$$w(t, x) = t|x|^2,$$

one sees that

$$w_t(t, x) = |x|^2,$$
$$w_x(t, x) = 2tx.$$

Thus the given problem is equivalent to

$$U_t + |U_x|^2 = 0,$$
$$U(0, x) = f(x).$$

Corollary 6.1 solves this problem, so the original problem has solution

$$u(t, x) = \min_y\{f(y) + |x - y|^2/4t\} - t|x|^2.$$

In particular, if $f = 0$, u is given by $u(t, x) = -t|x|^2 = -w.$ ∎

Transformation to an equivalent problem is often useful when one wishes to work with a nonnegative Lagrangian. As long as L is bounded below by l, one may consider the equivalent Lagrangian $L - l$, as shown in the following corollary. Examining the proof of lemma 7.3, one sees that such a transformation would have simplified matters slightly.

COROLLARY 8.1   Let (H, f) be an initial value problem, $L = H^*$ as before, g a continuous real function on $[b, \infty)$. Then $u: D \to \mathbb{R}$ solves the problem (H, f) if and only if $U = u + \int g$ solves the problem:

$$u_t + H(t, x, u_x) - g(t) = 0,$$

$$u|B = f + \int g.$$

Here $\int g$ denotes any indefinite integral of g and the new Lagrangian is

$$\mathscr{L}(t, x, q) = L(t, x, q) + g(t).$$

Proof   Simply let $w(t, x) = \int g$ in lemma 8.1.   ∎

Exercise

8.1   Prove lemma 8.1.

9. A LITTLE u DEPENDENCE?

For this section we generalize the Hamilton–Jacobi equation to

$$u_t + H(t, x, u, u_x) = 0. \tag{9.1}$$

That is, the Hamiltonian is allowed to depend also upon u. Why bother? Did we not claim in chapter I, section 9, that any first-order equation may be transformed into a Hamilton–Jacobi equation? Yes, *locally*! The implicit definition $V(x, u) = 0$ of u can only be solved for u locally in general. Thus where global solutions are needed, this approach breaks down. Even if it happens that V is such that u can be determined globally, it is very difficult to assign properties to the original equation which yield a V with the correct characteristics. Thus theorems are difficult to come by, although it never hurts to try this technique on a given problem.

Avron Douglis in [87] develops a very general treatment of the Cauchy problem where H is allowed to depend upon u. He proves existence, uniqueness (in a certain class of solutions), and continuous dependence upon initial data for a wide class of first-order equations. The only excuse for not presenting these results here is lack of space, and the reader with a Cauchy problem to analyze is strongly urged to consult Douglis' paper.

We will content ourselves with a slight extension of the variational method using a Picard iteration. This extension is for a special mixed problem, and is intended more to stimulate the imagination of the reader and to provoke further research than to present a general theorem.

THEOREM 9.1   *Third Existence Theorem*   Let $B = ([0, \infty) \times \partial K) \cup (\{0\} \times K)$, where K is a compact domain in $\mathbb{R}^n$ and ∂K its topological boundary, and consider the problem

$$u_t + H(t, x, u, u_x) = 0, \qquad t > 0, \quad x \in \text{int}(K),$$

$$u \,|\, B = f.$$

Suppose that H and f satisfy **(H1)–(H6)** and **(D1)–(D3)** of theorem 7.1, where (D2) is interpreted to mean

$$f(t, x) - f(s, y) \le \int_s^t L(\tau, \alpha(\tau), v(\tau, \alpha(\tau)), \alpha'(\tau)) \, d\tau$$

for any continuous v and, in addition,

   (H7) $|H_u| \le g_6(t)$, g_6 continuous.

Then there exists a global solution u to this problem.

Proof Choose $T > 0$ so that

$$\eta = \int_0^T g_6 < \tfrac{1}{2}$$

and temporarily restrict consideration to the compact set $\mathcal{K} = [0, T] \times K$. For any continuously differentiable function $u \colon \mathcal{K} \to \mathbb{R}$, define

$$\mathcal{T}u = \min_\alpha \left\{ f(s, y) + \int_s^t L(\tau, \alpha, u(\tau, \alpha), \alpha') \, d\tau \right\}$$

as in theorem 7.1.

Now let us establish that theorem 7.1 does apply to

$$\mathcal{H}(t, x, p) = H(t, x, u(t, x), p).$$

First note that **(H1)** follows by hypothesis, as do **(H3)**, **(H5)**, and **(H6)**. **(H2)** holds because H and u are assumed continuously differentiable. Now since **(H4)** holds for H,

$$|\mathcal{H}_x| \le g_1(t)(pH_p - H) + g_2(t) + |H_u| \cdot |u_x|$$
$$\le g_1(pH_p - H) + g_2 + g_6 M,$$

where M is a bound for $|u_x|$ on $\mathcal{K}$. Thus **(H4)** holds for $\mathcal{H}$ with $g_2 + Mg_6$ replacing g_2. **(D1)–(D3)** apply by hypothesis.

Thus by theorem 7.1, $v = \mathcal{T}u$ is locally Lipschitzian on $D = \text{int}(\mathcal{K})$, satisfies

$$v_t + \mathcal{H}(t, x, v_x) = 0 \qquad \text{a.e.}$$

and, since all boundary approaches are deterministic, defining $v = f$ on B, v is continuous on all of $\mathcal{K}$.

We would now like to define $\mathcal{T}^2 u$, $\mathcal{T}^3 u$, etc. and proceed á la Picard. However, for theorem 7.1 to apply at any given step, it is necessary that H, hence u, be smooth **(H2)**. Thus we must smooth $\mathcal{T}u$ before defining $\mathcal{T}^2 u$ etc. Mollifying functions are naturally used for this purpose. However, they require that v be defined and continuous in a neighborhood of $\mathcal{K}$ in order to meet our needs. But, since $\mathscr{K}$ is compact, we may more simply approximate by a polynomial.

So define $u_0 = 0$ and $u_1 = \mathscr{T} u_0$. Choose a smooth function (polynomial) v_1 on $\mathscr{K}$ so that

$$\|v_1 - u_1\|_\infty < \eta^2/2.$$

Continue this way, obtaining

$$v_0 = u_0, u_1, v_1, u_2, v_2, u_3, \ldots$$

where v_i is smooth,

$$\|v_i - u_i\|_\infty < \eta^{i+1}/2^i$$

and

$$u_{i+1} = \mathscr{T} v_i$$

Now if β is an extremal for defining u_1 from v_0,

$$u_2(t, x) - u_1(t, x) = \min_\alpha \left\{ f(s, y) + \int_\alpha L(\tau, \alpha, v_1, \alpha') \, d\tau \right\}$$

$$- \left\{ f + \int_\beta L(\tau, \beta, v_0, \beta') \, d\tau \right\}$$

$$\leq \int_\beta |L(\tau, \beta, v_1, \beta') - L(\tau, \beta, v_0, \beta')| \, d\tau.$$

Reversing the roles of u_1 and u_2, we obtain

$$\|u_2 - u_1\|_\infty \leq \eta \|v_1 - v_0\|_\infty$$

$$\leq \eta(\|v_1 - u_1\|_\infty + \|u_1 - u_0\|_\infty)$$

$$\leq \eta(\eta^2/2 + \|u_1 - u_0\|_\infty).$$

In general, since $\|v_i - u_i\|_\infty < \eta^{i+1}/2^i$ and $\eta < \frac{1}{2}$, we have

$$\|u_{m+1} - u_m\|_\infty < \eta r_m + \sum r_i \eta^{m-1-i} + \eta^m \|u_1 - u_0\|_\infty,$$

where $r_i = \eta^{i+1}/2^i$ and the sum is over $1 \leq i \leq m - 1$. This may be established easily by induction. Thus

$$\|u_{m+1} - u_m\|_\infty < \eta^m(1 + \|u_1 - u_0\|_\infty)$$

and

$$\|u_{m+k} - u_m\|_\infty < \eta^m(1 + \|u_1 - u_0\|_\infty)/(1 - \eta),$$

so that $\{u_k\}$ is a Cauchy sequence in $C(\mathscr{K})$.

Let $u_k \to u \in C(\mathcal{K})$ uniformly on $\mathcal{K}$. Clearly $v_k \to u$ uniformly too. Since L is continuous, for each curve α,

$$f + \int_\alpha L(\tau, \alpha, v_k, \alpha') \, d\tau \to f + \int_\alpha L(\tau, \alpha, u, \alpha') \, d\tau$$

as $k \to \infty$. That is, $\mathcal{T} v_k \to \mathcal{T} u$ pointwise, although theorem 7.1 does not necessarily apply to $\mathcal{T} u$. But $\mathcal{T} v_k = u_{k+1} \to u$ uniformly. Thus $\mathcal{T} u = u$ is a fixed point of $\mathcal{T}$. That is,

$$u(t, x) = \min_\alpha \left\{ f(s, y) + \int_s^t L(\tau, \alpha, u(\tau, \alpha), \alpha') \, d\tau \right\}. \tag{9.2}$$

Since u is continuous on $\mathcal{K}$, hence bounded, it is easy to repeat the proofs of section 5 to show that u is locally Lipschitzian on int($\mathcal{K}$). Since (9.2) holds, and $L(t, x, u(t, x), q)$ and $\mathcal{H}$ are continuous in t and x, theorem 3.1 applies and (9.1) holds wherever u is differentiable (almost everywhere), and has an extremal.

Now fix $(t, x) \in$ int($\mathcal{K}$). If α_k is an extremal for defining $u_k(t, x)$ from v_{k-1} once $\|u_k - u\|_\infty$ is sufficiently small, it is easy to show, as in the proof of lemma 7.3, that $\|\alpha_k'\|_\infty$ is bounded by a constant r independent of k. Thus without loss of generality, some subsequence, $\{\alpha_k\}$ must converge uniformly to a Lipschitzian curve α. Then

$$f(s_k, y_k) + \int_{\alpha_k} L = u_k(t, x) \to u(t, x)$$

but by lower semicontinuity,

$$u(t, x) \le f(s, y) + \int_\alpha L \le \liminf u_k(t, x).$$

Thus α is an extremal for computing $\mathcal{T} u$. Thus each point has an extremal, so that u satisfies (9.1) almost everywhere in int($\mathcal{K}$).

So u is a global solution on $\mathcal{K}$. Now consider $(t, x) \in B$ for $t \in [T, T_1]$, where $T_1 > T$ and $\int_T^{T_1} g_6 < \frac{1}{2}$. Redefining the problem using $\{T\} \times K$ in place of $\{0\} \times K$ for part of the boundary and $u(T, x)$ for the data there, the same process applies to $[T, T_1]$ as long as (D2) can be reestablished for the new problem. But for any continuous v and any curve $\alpha : [s, t] \to \mathbb{R}^n$, (D2) holds by hypothesis if $s > T$ (the original data is involved). If $s = T$,

$$f(t, x) - u(T, w) = f(t, x) - f(s, y) - \int_s^T L(\tau, \beta, u, \beta') \, d\tau,$$

where β is an extremal for defining $u(T, w)$. By the original hypothesis (D2) applied along α then β,

$$f(t, x) - u(T, w) \leq \left(\int_{\beta} L(\tau, \beta, u, \beta') \, d\tau + \int_{\alpha} L(\tau, \alpha, v, \alpha') \, d\tau \right)$$

$$- \int_{s}^{T} L(\tau, \beta, u, \beta') \, d\tau$$

$$= \int_{\alpha} L(\tau, \alpha, v, \alpha') \, d\tau$$

as was desired. Reapplying the Picard-type of iteration scheme, we may extend u to $[T, T_1] \times K$, thus to $[0, T_1] \times K$ as a global solution. Note that, although u is locally Lipschitzian for $x \in \text{int}(K)$ and $t \in (0, T)$ or $t \in (T, T_1)$, u might not be locally Lipschitzian for $t = T$. However, it is easy to show that u is locally Lipschitzian on all of $(0, T_1) \times \text{int}(K)$ either directly as in previous proofs or by varying T slightly (keeping $\eta < \frac{1}{2}$), and seeing that the same u results, so that $t = T$ presents no special problems.

Finally any finite interval $[0, M]$ may be broken into m intervals $I_k = [T_{k-1}, T_k]$, where

$$0 = T_0 < T_1 < \cdots < T_m = M,$$

and

$$\int_{I_k} g_6 < \tfrac{1}{2}, \qquad 0 \leq k \leq m,$$

so u may be defined inductively on $[0, M]$. Thus u can be extended on all of $[0, \infty) \times K$ and yields the desired global solution. ∎

Exercises

9.1   Generalize the third existence theorem. When you do, publish it.

9.2   Use the transformation (I.9.1) and (I.9.2) to produce global existence theorems. Publish them.

10. OTHER EXISTENCE TECHNIQUES

The existence theorems of this chapter are among the most general now known, although no attempt at maximum generality has been made. The technique used was an extension of the classical variational method, requiring global convexity and absolute minimums, as opposed to the local classical theory. However, there are many existence techniques known, some classical and some more recent. Almost all of the known techniques have been applied to the Hamilton–Jacobi equation at some time or other, with varying degrees of success. This last section on existence attempts to summarize the results of applying the various other methods to obtain global solutions.

A very common technique in partial differential equations is the approximation of solutions by a finite difference scheme. In the limit this often proves existence of solutions, and it also offers a useful computational device. In Oleinik's landmark paper [203], a difference scheme of Lax was used to obtain a solution to the one-dimensional Cauchy problem. Douglis [87] used a combination differential-difference scheme, and obtained a very general existence theorem for the n-dimensional Cauchy problem, in which H is allowed to depend upon u, as well as t, x, and u_x.

In Conway and Hopf [73], the variational approach was used for one-dimensional space-time independent Hamilton–Jacobi equations, but was immediately shown to be equivalent to an extension of the classical complete integral techniques. That is, if

$$u(t, x) = \min_{(s, y)} \{ f(s, y) + (t - s)L((x - y)/(t - s)) \},$$

this may be expanded to:

$$u(t, x) = \min_{(s, y)} \max_{p} \{ f(s, y) + p(x - y) - (t - s)H(p) \},$$

where the quantity in brackets is a complete integral. Hopf [144] and Aizawa and Kikuchi [3] extended this technique to the n-dimensional problem and in some cases where f is convex, were able to reverse the above min and max to obtain solutions with minimal hypotheses on H. The key to this extension of the complete integral techniques is a simple but powerful envelope lemma of Hopf [144]. We present the following theorem of Hopf here because it is the only global existence result known to the author for nonconvex H.

THEOREM 10.1 *Fourth Existence Theorem* (*Hopf*) Consider the n-dimensional Cauchy problem

$$u_t + H(u_x) = 0, \qquad t \geq 0, \qquad x \in \mathbb{R}^n,$$

$$u(0, x) = f(x).$$

If H is continuous and f is convex and Lipschitzian, then

$$u(t, x) = \max_{p} \min_{y} \{ f(y) + p(x - y) - tH(p) \}$$

solves the problem (H, f). Furthermore u is Lipschitzian in x with the same Lipschitz constant as f for all $t > 0$.

Proof See Hopf [144, pp. 966–971]. ∎

COROLLARY If "max" and "min" are interchanged in theorem 10.1, and "convex" replaced by "concave," the theorem still holds.

Proof Exercise 10.1. ∎

The variational technique was pushed to yield quite general existence of Cauchy problem solutions by Fleming [109]. In this paper, it was also shown that the variational solution is the same as that obtained as a limit as $\epsilon \to 0$ of solutions of the higher (second) order parabolic equations

$$u_t + H(t, x, u) = \epsilon \Delta_x u,$$

where Δ_x denotes the Laplace operator in x, assuming suitable restrictions on H and f. This limit of solutions of higher order equations was also used by Oleinik [203] for the one-dimensional case and by Hopf [143] for Burger's equation.

The last few years have seen some quite novel approaches. The most successful classical technique is the method of characteristics. It is natural to look to an extension of this technique. In Douglis [89] the extension is provided. The idea is to obtain strict solutions locally, that is, in a thin layer, then repeat the process for an adjoining layer, using the edge of the first layer as a "boundary." Continuing in this manner, smoothing and taking successively finer layers, a solution of the original problem is obtained in the limit. This process is very intuitive but rather complex in application.

Elliott and Kalton [98] and [99] show the existence of solutions to certain boundary value problems using game-theoretic techniques. That is, they construct a differential game related to the given problem, use it to obtain a solution, then show that the solution obtained is independent of the choice of the related differential game.

Since the Hamilton–Jacobi equation is essentially an evolution equation, the methods of semigroups apply in some cases. See Goldstein [120] for a brief introduction to these methods. The work of Crandall and Liggett [80], among others, yields semigroups for nonlinear operators, such as that involved in the Hamilton–Jacobi equation. The general solutions obtained are defined differently than those obtained in this work since the derivative u_t is generally some type of strong derivative (limit in a space of functions of x). Aizawa [2] treats a one-dimensional Hamilton–Jacobi equation via semigroups and Burch [50, 51] develops a very nice n-dimensional treatment. A very readable summary of Burch's work is contained in Goldstein [121].

With these many techniques now available, it is likely that the near future will yield quite general existence theorems for general first-order equations and possibly a unified approach. The variational technique of this chapter was selected for its simplicity and the generality of the boundaries which can be considered. However the variational technique seems intrinsically limited to the Hamilton–Jacobi equation, while other techniques, such as the layering methods, seem capable of unlimited extension.

Exercises

10.1   Prove the corollary to theorem 10.1 from the theorem.

10.2   Prove existence theorems by techniques other than the variational approach. For a Cauchy problem this will probably require a review of the literature. For the general boundary value problem, this is strictly a research problem.

III Uniqueness and Properties of Solutions

PREAMBLE

In most physical situations it is not enough to find a solution; if it is to be used to predict behavior it must be shown to be the only solution. Otherwise the model is incomplete, so further information is needed to restrict consideration to the physically meaningful result. For example, if a variational minimum is sought and the corresponding Hamilton–Jacobi equation has been solved, say by characteristic theory, it can not be claimed that the solution represents the variational minimum unless a uniqueness theorem precludes extraneous solutions.

In addition to uniqueness, it is usually necessary that the solution depend continuously upon the boundary data, and perhaps upon the PDE itself since these are generally measured quantities which might err somewhat. In other words, the problem should be, in the mathematical sense, well posed.

For the Hamilton–Jacobi equation, particularly in dealing with global solutions, these questions are intricate and involve numerous subtleties. As might be expected, not all of the answers are known but the Cauchy problem is fairly well understood. Kruzkov [163] and Douglis [87] have shown independently that the Cauchy problem has a unique solution in a suitably restricted class of solutions and Feltus [101] has shown uniqueness in a similar class of solutions for mixed problems.

In this chapter we shall present a few simple properties of the variational solution, including the semiconcavity property needed for uniqueness, and then study some of the results of Douglis and Feltus. We will also show that the variational solution is always a maximal solution.

1. BASIC INEQUALITIES

The variational solution is relatively easy to study because of its representation as an integral along an extremal. To illustrate this point, recall the proof
that the variational solution is locally Lipschitzian. We will begin here by
noting a few inequalities that follow directly from the variational representation. With the notation and terminology of Chapter II, let u be the variational
solution and (t, x) a point of D.

LEMMA 1.1   If (s, y) is any point of B, with $s < t$, then

$$u(t, x) \le f(s, y) + \int_s^t L(\tau, y + (\tau - s)a, a)\, d\tau, \tag{1.1}$$

where $a = (x - y)/(t - s)$.
   If $b > -\infty$ and (b, x) is a boundary point,

$$u(t, x) \le f(b, x) + \int_b^t L(\tau, x, 0)\, d\tau. \tag{1.2}$$

Proof   For (1.1), $u(t, x)$ cannot exceed the value obtained by integrating
along the line segment joining (t, x) and (s, y). For (1.2), join (t, x) and (b, x)
by a line segment.   ∎

We will now obtain a very useful inequality by deforming an extremal α
joining (t, x) and its foot point (s, y). If (S, Y) is another boundary point and
(T, X) another point of D, $S < T$, define the curve β joining (T, X) and (S, Y)
by

$$\beta(\sigma) = \alpha(\tau) + (T - \sigma)(Y - y)/(T - S) + (\sigma - S)(X - x)/(T - S), \tag{1.3}$$

where

$$\tau = s + (\sigma - S)(t - s)/(T - S)$$
$$= s(T - \sigma)/(T - S) + t(\sigma - S)/(T - S).$$

Note that

$$d\tau/d\sigma = (t - s)/(T - S)$$

and

$$\beta'(\sigma) = \alpha'(\tau)(t - s)/(T - S) + (X - x - Y + y)/(T - S).$$

The curve β has a shape very similar to α but, of course, can not be expected
to be an extremal. However, for small deformations, β is close to an extremal
and very useful inequalities come from integrating along β. That is, they are
implied by the fact that

$$u(T, X) \le f(S, Y) + \int_\beta L.$$

Written out in detail, this becomes

$$u(T, X) - f(S, Y)$$

$$\leq \int_S^T L\left(\sigma, \alpha(\tau) + \frac{(T - \sigma)(Y - y) + (\sigma - S)(X - x)}{T - S},\right.$$

$$\left.\frac{(t - s)\alpha'(\tau) + (X - x - Y + y)}{T - S}\right) d\sigma$$

$$= \frac{T - S}{t - s} \int_s^t L\left(\sigma, \alpha(\tau) + \frac{(T - \sigma)(Y - y) + (\sigma - S)(X - x)}{T - S},\right.$$

$$\left.\frac{(t - s)\alpha'(\tau) + (X - x - Y + y)}{T - S}\right) d\tau. \qquad (1.4)$$

LEMMA 1.2   In each case following the given inequality is valid if it makes sense:

$$u(t, z) - u(t, x) \leq \int_s^t \left\{ L\left(\tau, \alpha + \frac{(\tau - s)(z - x)}{t - s}, \alpha' + \frac{z - x}{t - s}\right)\right.$$

$$\left. - L(\tau, \alpha, \alpha') \right\} d\tau \qquad (1.5)$$

$$u(r, x) - u(t, x) \leq \int_s^t \{(r - s)(t - s)^{-1} L(s + (\tau - s)$$

$$\times (r - s)/(t - s), \alpha, (t - s)\alpha'/(r - s)) - L(\tau, \alpha, \alpha')\} d\tau. \qquad (1.6)$$

$$u(t, z) - u(t, x) \leq f(s, y + z - x) - f(s, y)$$

$$+ \int_s^t \{L(\tau, \alpha + z - x, \alpha') - L(\tau, \alpha, \alpha')\} d\tau. \qquad (1.7)$$

$$u(r, x) - u(t, x) \leq f(s + r - t, y) - f(s, y)$$

$$+ \int_s^t \{L(\tau + r - t, \alpha, \alpha') - L(\tau, \alpha, \alpha')\} d\tau. \qquad (1.8)$$

$$u(t, z) - u(t, x) \leq f(t - (t - s)|z - y|/|x - y|, y) - f(s, y)$$

$$+ \int_s^t \left\{\frac{|z - y|}{|x - y|} L\left(\sigma, \alpha + (\tau - s)\right.\right.$$

$$\left.\left. \times (z - x)/(t - s), \frac{|x - y|}{|z - y|}\left(\alpha' + \frac{z - x}{t - s}\right)\right) - L(\tau, \alpha, \alpha')\right\} d\tau,$$

$$\qquad (1.9)$$

where $\sigma = t - (t - \tau)|z - y|/|x - y|$.

$$u(r\ x) - u(t, x) \le f(s, y + (t - r)(x - y)/(t - s)) - f(s, y)$$

$$+ \int_s^t \left\{ \frac{(r - s)}{(t - s)} L\left(\sigma, \alpha + \frac{(r - \sigma)}{(r - s)} \frac{[x(t - r) + y(r - t)]}{(t - s)}, \frac{(t - s)\alpha'}{(r - s)} \right. \right.$$

$$\left. \left. - \frac{[x(t - r) + y(r - t)]}{(r - s)(t - s)} \right) - L(\tau, \alpha, \alpha') \right\} d\tau, \tag{1.10}$$

where $\sigma = s + (\tau - s)(r - s)/(t - s)$.

Proof   For (1.5) let $(T, X) = (t, z)$, $(S, Y) = (s, y)$ in (1.4). For (1.6), let $T = r$ and s, x, and y remain unchanged. For (1.7) simply translate α by letting $T = t$, $S = s$, $Y = y + (z - x)$. For (1.8), let $X = x$, $Y = y$, $S = s + r - t$. To obtain (1.9), after translating (t, x) to $(T, X) = (t, z)$, let $Y = y$ and determine S by making $a = (x - y)/(t - s) = (z - y)/(t - S)$. That is,

$$S = t - (t - s)|z - y|/|x - y|.$$

For (1.10), let $S = s$, $X = x$, r be given, and determine Y by $a = (x - y)/(t - s) = (x - Y)/(r - s)$ or

$$Y = x - (r - s)(x - y)/(t - s). \qquad \blacksquare$$

Note that (1.7) and (1.10) are useful for Cauchy problems since $S = s = b$ in that case. Similarly (1.8) and (1.9) are applicable to classical boundary value problems since $Y = y$. Finally (1.5) and (1.6) are valid for any boundary since $(S, Y) = (s, y)$. Many similar inequalities may be obtained by choosing various combinations of boundary points (S, Y), curves β, and line segments, but lemmas 1.1 and 1.2 yield most of those needed in our discussion.

In case H is independent of t and x, these lemmas yield many interesting properties of u very simply. Starting with the Cauchy problem, we have the following theorem.

THEOREM 1.1   Let B be all of $\{b\} \times \mathbb{R}^n$, $D = (b, \infty) \times \mathbb{R}^n$, $f \in C(\mathbb{R}^n)$, and H a real convex function on $\mathbb{R}^n$ such that $H(p)/|p| \to \infty$ $|p| \to \infty$. Suppose also that f is bounded below, say by $-M$ where $M > 0$. Let L be the convex dual of H and let l be the minimum value of L. Then

$$u(t, x) = \min_y \{ f(y) + (t - b)L((x - y)/(t - b)) \}$$

has the following properties:

(1) u is locally Lipschitzian on D, hence differentiable almost everywhere.
(2)   For all (t, x) where u is differentiable,

$$u_t(t, x) + H(u_x(t, x)) = 0.$$

(3) $u \cup f$ is continuous on $B \cup D$. More precisely, if $(t_k, x_k) \to (b, x)$ is any boundary approach, then

$$u(t_k, x_k) \to f(x).$$

(4)   If f is Lipschitzian, then there is a constant $Q > 0$ such that each $(t, x) \in D$ has a foot point (b, y) with

$$|a| = |x - y|/(t - b) \le Q.$$

(5)   If f has Lipschitz constant K, then u is Lipschitzian and, for each $t > b$, has x-Lipschitz constant K.

(6)   If f is also bounded above by M, then

$$|u(t, x) - l(t - b)| \le M.$$

Thus u remains bounded as long as t remains bounded. In particular, if $l = 0$, then u is bounded.

(7)   If f is semiconcave with constant C, then for each t, u is semiconcave in x with constant C.

(8)   If f is bounded, then for any $\epsilon > 0$ there is a constant Q such that each (t, x) with $t \ge b + \epsilon$ has a foot point (b, y) with

$$|a| = |x - y|/(t - b) \le Q.$$

(9)   If f is bounded and H is strictly convex, then for any $\epsilon > 0$, u is Lipschitzian for $t \ge b + \epsilon$.

(10)   If L is semiconcave with constant C, then for each t, u is semiconcave in x with constant $C/(t - b)$.

Proof   (1)–(3) are merely a repetition of the first existence theorem. For (4), let $a = (x - y)/(t - b)$, where (b, y) is a foot point for (t, x). Then

$$u(t, x) = f(x - (t - b)a) + (t - b)L(a)$$

$$\le f(x) + (t - b)L(0)$$

by (1.2). This yields

$$L(a)/|a| \le [f(x) - f(x - (t - b)a)]/(t - b)|a| + L(0)/|a|$$

$$\le K + L(0)/|a|,$$

where K is a Lipschitz constant for f. The right-hand side here is bounded as a function of a for large $|a|$, but as $|a| \to \infty$, $L(a)/|a| \to \infty$, so that $|a|$ must remain bounded, say by $Q > 0$.

Now suppose that f is Lipschitzian with constant K. Then by (1.7),

$$u(t, z) - u(t, x) \le f(y + z - x) - f(y)$$

$$\le K|z - x|,$$

where (b, y) is a foot point for (t, x). By symmetry, K is an x-Lipschitz constant for u. Now if $r > t$, integrate along an extremal to (t, x), then along the straight line segment from (t, x) to (r, x) to calculate that

$$u(r, x) - u(t, x) \le (r - t)L(0).$$

On the other hand, if $r < t$, let $(r, \alpha(r))$ replace (t, x) in (1.7), and use (r, x) instead of (t, z), to discover another general inequality:

$$u(r, x) - u(t, x) \le f(s, y + x - \alpha(r)) - f(s, y) - \int_r^t L(\tau, \alpha, \alpha') + \int_s^r \Delta L \, d\tau,$$

(1.11)

where

$$\Delta L = L(\tau, \alpha + x - \alpha(r), \alpha') - L(\tau, \alpha, \alpha').$$

In our case, L is independent of t and x, so

$$u(r, x) - u(t, x) \le f(x - (r - b)a) - f(x - (t - b)a) - (t - r)L(a),$$

where $\alpha'(\tau) \equiv a$. Combining these two cases, if $t > r$,

$$-(t - r)L(0) \le u(r, x) - u(t, x) \le K(t - r)|a| - (t - r)L(a).$$

By part **(4)**, a above is bounded, and **(5)** is established.

For **(6)** simply note that

$$- M + l(t - b) \le u(t, x)$$

$$\le f(x - (t - b)a) + (t - b)L(a)$$

for any $a \in \mathbb{R}^n$. In particular, for any a which minimizes L,

$$u(t, x) \le M + (t - b)l.$$

To conclude **(7)** above, let (b, y) be a foot point for (t, x) and use (1.7) twice, to see that

$$[u(t, x + h) - u(t, x)] + [u(t, x - h) - u(t, x)]$$

$$\le [f(y + h) - f(y)] + [f(y - h) - f(y)]$$

$$\le C|h|^2.$$

Part **(8)** is proved as was part **(4)**, except that now

$$L(a)/|a| \le 2M/(t - b)|a| + L(0)/|a|$$

$$\le 2M/|a|\epsilon + L(0)/|a|.$$

For **(9)**, note that H strictly convex implies that L is continuously differentiable, so that $|L_q|$ is bounded as long as $|q|$ is bounded, and by part **(8)**,

$q = a = (x - y)/(t - b)$ where (b, y) is a foot point is bounded for $t \geq \epsilon$. That is, for the arguments considered, L is Lipschitzian, say with constant N. Then by (1.5),

$$u(t, z) - u(t, x) \leq (t - b)[L((z - y)/(t - b)) - L((x - y)/(t - b))]$$
$$\leq N|z - x|$$

for the proper y. So u is Lipschitzian in x for $t \geq b + \epsilon$. Now for the time argument, by (1.6),

$$u(r, x) - u(t, x) \leq (t - s)(r - s)(t - s)^{-1}[L((t - s)a/(r - s)) - L(a)]$$
$$\leq |r - b|N|a|((t - b)/(r - b) - 1)$$
$$= N|a||t - r|$$
$$\leq NQ|t - r|.$$

Finally for (10) apply (1.5) to yield

$$u(t, x + h) - u(t, x) \leq (t - b)[L((x + h - y)/(t - b)) - L((x - y)/(t - b)),$$

where again (b, y) is a foot point for (t, x). Applying (1.5) again to $(t, x - h)$ and (t, x) results in

$$[u(t, x + h) - u(t, x)] + [u(t, x - h) - u(t, x)]$$
$$\leq (t - b)[C|h|^2/(t - b)^2]$$
$$= [C/(t - b)]|h|^2. \quad \blacksquare$$

COROLLARY 1.1   If, in theorem 1.1, B is a proper subset of $\{b\} \times \mathbb{R}^n$, then parts (1), (2), and (10) of the theorem still hold, but parts (3)–(7) are false in every case. Parts (8) and (9) are valid if and only if the complement of B in $\{b\} \times \mathbb{R}^n$ is bounded.

Proof   Exercise 1.1. $\blacksquare$

For general boundaries, the results of theorem 1.1 do not generally hold. Of course (1) and (2) are part of the first existence theorem and (3) holds for all deterministic approaches. But if b is finite, but B_b is not all of $\{b\} \times \mathbb{R}^n$, there are nondeterministic approaches and (3)–(7) are false. However, for specific boundaries, or for suitably restricted domains D, (1.1)–(1.11) still yield useful information. We will present here a few of these special cases, and leave the proofs for exercises.

THEOREM 1.2   Let $B = \mathbb{R} \times X$, with $X \subseteq \mathbb{R}^n$. If (H, f) is a boundary value problem satisfying the hypotheses of the first existence theorem with boundary B, the (1)–(4) of theorem 1.1 hold. Also

(5)   If f has Lipschitz constant K, then u has t-Lipschitz constant K. If L is also continuously differentiable, then u is Lipschitzian.

(6)   If, in addition to (D3) of the first existence theorem, f grows sufficiently fast so that $f(s, y) + (\epsilon - 2l)s \to \infty$ as $s \to -\infty$ uniformly in y for some $\epsilon > 0$, then each point has an extremal α with $\alpha'(\tau) \equiv a$ satisfying $\epsilon/N \le |a| \le Q$ for some $N, Q > 0$. That is, $|a|$ is also bounded away from zero. Note that this always holds by (D3) if $l > 0$.

(7)   In (6) above, if f is Lipschitzian, then so is u (even if L is not of class C^1).

(8)   If L is semiconcave with constant C, then for (t, x) with distance $(x, X) \ge d > 0$, u is semiconcave in x with constant CQ/d, where Q is a bound on $|a| = \|\alpha'\|_\infty$ for extremals. In particular, if f is concave, u is globally concave.

(9)   If f is semiconcave with constant C, then u is semiconcave in t with constant C. In case X is a hyperplane and the hypothesis of (6) holds, u is uniformly semiconcave in x on either side of $\mathbb{R} \times X$. Whenever X is a hyperplane, u is semiconcave with constant C in any hyperplane parallel to $\mathbb{R} \times X$.

Proof   Exercise 1.2.   ∎

COROLLARY 1.2   ·If $B = [b, \infty) \times X$, the results of theorem 1.2 hold if D is restricted to points (t, x) with

$$\text{distance}(x, X) \le (t - b)M$$

for some $M > 0$.

Proof   Exercise 1.3.   ∎

For more complicated boundaries, these results may be applied piecewise, considering only part of the boundary at a time. Then, hopefully, near each piece of the boundary the solution will depend only upon that piece of the boundary. Feltus, in [101], establishes properties of the variational solution for a one-dimensional mixed problem by similar methods. His results include establishing the uniform semiconcavity condition which we will later see is useful for establishing uniqueness.

In the case of a general Hamiltonian $H(t, x, p)$, the corresponding results are more complicated, but may still be established by using (1.1)–(1.11) and suitably bounding $|L_x|$. We will consider only the Cauchy problem with $b = 0$.

THEOREM 1.3   Consider the Cauchy problem (H, f), where $B = \{0\} \times \mathbb{R}^n$ and H satisfies (H1)–(H6) of the second existence theorem, with (H3) strengthened to

(H3′) $H(t, x, p)/|p| \geq g_6(|p|)$,

where

$$g_6(r) \nearrow \infty \qquad \text{as} \quad r \to \infty$$

and (H5) strengthened to

(H5′) $g_3(t) \leq pH_p - H \leq g_7(t, p)$,

where g_7 is continuous.
    Suppose that f satisfies:

(D1′) f is of class C^1.
(D3′) $|f|, |f_x| \leq K$, for some $K > 0$.

Then in any strip $0 \leq t \leq T$, the variational solution, $u(t, x)$, is uniformly Lipschitzian. Here $u(0, x)$ is defined to be $f(x)$.

    Proof   Note first, that since all t considered lie in $[0, T], g_1, g_2, g_3$ may be considered to be constant. Also by (H3′),

$$L(t, x, 0) \leq -g_6(0).$$

Thus by (1.2),

$$-K^* - T|g_3| \leq u(t, x)$$

$$\leq f(x) + \int_0^t L(\tau, x, 0)\, d\tau$$

$$\leq K + T|g_6(0)|.$$

Thus u is bounded, say by $M_1 > 0$. Also, along any extremal α, joining (t, x) with its foot point $(0, y)$,

$$\int_0^t |L_x(\tau, \alpha, \alpha')|\, d\tau \leq g_1 \int_\alpha L + g_2$$

$$\leq g_1 |u(t, x) - f(y)| + g_2$$

$$\leq g_1(M_1 + K) + g_2$$

$$\leq M_2$$

for some $M_2 > 0$.

Using the maximum principle again, we have for almost all $\tau \in (0, t)$,

$$qh(\tau) - L(\tau, \alpha(\tau), q)$$

is maximized over all $q \in \mathbb{R}^n$, at $q = \alpha'(\tau)$, where now

$$h(\tau) = f_y(y) + \int_\alpha L_x$$

$$\leq K + M_2.$$

Thus

$$\alpha'(\tau) = H_p(\tau, \alpha, h(\tau)),$$

so

$$|\alpha'| \leq g_4(K + M_2).$$

Hence $|\alpha'|$ is bounded independent of (t, x), say by $M_3 > 0$.

Now by **(H5')**, L is bounded above for $0 \leq t \leq T$ and $|q| \leq M_3$, independent of x. Thus by **(H4)**, so is $|L_x|$, say by M_4. Therefore (1.7) yields

$$u(t, z) - u(t, x) \leq K|z - x| + \int_0^t L_x(\tau, \alpha + \theta(z - x), \alpha')(z - x)\, d\tau$$

$$\leq K|z - x| + tM_4|z - x|$$

$$\leq (K + TM_4)|z - x|$$

$$= M_5|z - x|,$$

where $0 < \theta < 1$. That is, u is Lipschitzian in x.

Finally for $t > r$, a slight variation of (1.11) yields

$$u(t, x) - u(r, \alpha(r)) = \int_r^t L(\tau, \alpha, \alpha')\, d\tau,$$

so that

$$|u(t, x) - u(r, x)| \leq \int_r^t L + M_5|x - \alpha(r)|$$

$$\leq (t - r)\sup|L| + M_3 M_5(t - r),$$

where the supremum is over all t, x and $|q| \leq g_6(M_3)$. ∎

The above proof is essentially that of Fleming [109, pp. 515–519]. The semiconcavity of u in x no longer follows trivially from that of f or L as in theorem 1.1, because of the influence of the x dependence of L on (1.7) and

(1.5). However, if L_{xx} is suitably bounded, similar results are still possible. For example:

THEOREM 1.4   Consider the Cauchy problem of theorem 1.3.

(A)   If f is semiconcave with constant C and L is semiconcave in x with constant $\lambda(t, q)$, where λ is continuous, then in any layer $0 \leq t \leq T$, u is semiconcave in x, uniformly in t.

(B)   If L is semiconcave in x and q simultaneously with constant $\lambda(t, q)$, that is, for $P = (x', q') \in \mathbb{R}^{2n}$,

$$\Delta^2 L(t, x, q) = L(t, (x, q) + P) + L(t, (x, q) - P) - 2L(t, x, q)$$
$$\leq \lambda(t, q)|P|^2,$$

where λ is continuous, then for $0 < t \leq T$, u is semiconcave in x with a constant of the form $K(T + 1/t)$.

Proof   For **(A)** simply use (1.7) twice and integrate. For **(B)**, use (1.5) twice and integrate.   ∎

In most cases it is desirable to have conditions upon H rather than upon L, to determine the properties of the solution. The easiest way is to require that H be a function of class C^2. Then $L_{qq} = H_{pp}^{-1}$ and semiconcavity is equivalent to:

$$h^T L_{qq} h \leq C|h|^2$$

for all $h \in \mathbb{R}^n$. In fact, if for any nonzero $h \in \mathbb{R}^n$,

$$0 < c_1|h|^2 \leq h^T H_{pp} h \leq c_2|h|^2$$

for the arguments involved, then letting $z = H_{pp}h$ or $h = L_{qq}z$,

$$z^T L_{qq} z \leq c_2|H_{pp}^{-1}z|^2$$

and

$$c_1|H_{pp}^{-1}z|^2 \leq h^T z \leq |H_{pp}^{-1}z| \cdot |z|$$

so that

$$|H_{pp}^{-1}z| \leq c_1^{-1}|z|$$

and

$$z^T L_{qq} z \leq c_2(c_1^{-1})^2|z|^2.$$

Fleming [109] established the following theorem, which we present without proof, by considering the theory of second-order parabolic equations and stochastic processes.

THEOREM 1.5   Consider the Cauchy problem (H, f) where $B = \{0\} \times \mathbb{R}^n$, f is continuously differentiable, bounded, and Lipschitzian, and H is of class C^3 and satisfies:

(1) $|H_t| + |H_p| + |H_{xp}| + |H_{tp}| \le C(|p|)$,

$$\gamma(|p|)|h|^2 \le h^T H_{pp} h \le C(|p|)|h|^2$$

for all $h \in \mathbb{R}^n$, where $\gamma(r)$ and $C(r)$ are positive functions, respectively non-increasing and nondecreasing in r.

(2) $H(t, x, p)/|p|$ as $|p| \to \infty$.

Given $r > 0$, there exists k_r such that $|H_p| \le r$ implies $|p| \le r$.

(3) $H(t, x, p) \ge M$ for some M.
(4) $|H_x| \le C_1(pH_p - H) + C_2$

for some positive C_1, C_2.

The variational solution is a Lipschitzian solution of this problem, for $0 < t \le T$, and is semiconcave in x with constant K/t for some K.   ∎

Fleming's theorem is about the best in this area, but the following easy special case may be established by our variational methods, so will be presented for instructional purposes.

THEOREM 1.6   Consider the Cauchy problem of theorem 1.3. If H also satisfies:

(1) H is of class C^2.
(2) $|H_{xx}| \le C_1$ for some C_1.
(3) $C_2|h|^2 \le h^T H_{pp} h \le C_3|h|^2$

for all $h \in \mathbb{R}^n$ (some $C_2, C_3 > 0$).

Then the result of theorem 1.5 is valid.

Proof   See exercise 1.4.   ∎

Exercises

1.1   Establish corollary 1.1. *Hint*: Let $(t, x) \to (b, x) \notin B$ to show that (3)–(7) are false.

1.2   Prove theorem 1.2. *Hint*: For (4), let $|q| \ge M \Rightarrow L(q) \ge 0$, and assume $|a| \ge M$. For (5) use (1.8) and (1.5). For (6), consider an equivalent problem where $l = \epsilon > 0$, and note that $L(a)/|a|$ and $|a|$ are bounded.

For (7) use (1.9). For (8) use (1.5). Part (9) follows from (1.8), (1.7), and finally an inequality like (1.9) where $(t, x + h)$ and $(t, x - h)$ are joined to B by straight line segments with $\alpha' = a = (x - y)/(t - s)$, where (s, y) is a foot point for (t, x).

1.3 Prove corollary 1.2. *Hint*: Extend B to $\mathbb{R} \times X$ by defining $f(s, y) = f(b, y) + N(b - s)$ for $s < b$, some $N > |l|$. Show that for some M and D as in corollary 1.2, the new variational solution coincides with that for $[b, \infty) \times X$. Letting N become larger, make M as desired.

1.4 Show that theorem 1.6 holds. *Hint*: Use (1.5) twice, add and subtract like quantities so as to consider x-differences in the integrands and q-differences separately. Then apply Taylor's theorem and the assumed bounds, to obtain a constant of semiconcavity of the form $K_1/t + K_2$, which is less than or equal to $(K_1 + K_2 T)/t$.

2. UNIQUENESS FOR THE CAUCHY PROBLEM

We are now in a position to consider the uniqueness problem. For the remainder of this section, the boundary B will be $\{0\} \times \mathbb{R}^n$. To squelch quickly any unfounded optimism, let us start with a nonuniqueness example. Several similar examples may be found in the works of Conway, Hopf, and Douglis. This particular example was suggested to the author by E. D. Conway.

Example 2.1   Consider the one-dimensional Hamilton–Jacobi equation:

$$u_t + u_x^2 = 0.$$

Three global solutions of this problem are

$$u_1(t, x) = 0,$$

$$u_2(t, x) = x - t,$$

$$u_3(t, x) = -x - t.$$

Since the minimum and maximum of any two solutions are again solutions, we have further solutions

$$u_4(t, x) = u_2 \vee u_3 = |x| - t,$$

$$u_5(t, x) = u_1 \wedge u_4 = \begin{cases} |x| - t, & |x| \leq t, \\ 0, & |x| \geq t. \end{cases}$$

Thus u_1 and u_5 are two different solutions which have the same Cauchy data. That is,

$$u_1(0, x) \equiv u_5(0, x) \equiv 0.$$

Note however that u_1 is semiconcave (actually concave), while u_5 is not semiconcave in x.

It turns out that the semiconcavity condition is what is needed to single out the "correct" solution. This condition may be considered a multi-dimensional generalization of the "entropy condition" encountered in the study of "shock waves" in two-dimensional quasilinear equations.

It has been noted that the "characteristic slope" or derivative of the extremal path α' is bounded independent of its endpoints, if proper conditions are placed upon H and f. Thinking of data as being carried along the extremal to determine the solution at its endpoints, it seems likely that the solution at (t, x) is determined completely by the data on the base of a cone with sufficiently steep sides:

$$\mathscr{B} = \{(0, y) : |y - x| \leq Mt\}.$$

This is the case, so we shall define cones of determinacy and show that the solution in such a cone is determined by the data on the base.

For the remainder of this section, we will follow the development of Douglis [87, pp. 1–10], except that we will consider only the Hamilton–Jacobi equation as defined throughout this work. Douglis actually obtained a more general theorem, allowing H to depend upon u as well as upon t, x, and p. It will be assumed only that H is convex and of class C^2. Thus for all t, x, and p and for all $h \in \mathbb{R}^n$,

$$h^T H_{pp} h \geq 0.$$

All solutions will be required to be uniformly Lipschitzian in any layer $S_T = [0, T] \times \mathbb{R}^n$, and to be semiconcave in x with constant $A/t + C$ for some positive A and C.

Now suppose that $|u_x| \leq P$ in S_T, and let K be the supremum of the "characteristic slopes," $|\alpha'| = |H_p(t, x, u_x)|$. More precisely, define

$$K = \sup\{|H_p(t, x, p)| : (t, x) \in S_T, |p| \leq P\}.$$

Define a *cone of determinacy* to be any cone of the form

$$D_{z, T} = \{(t, x) : 0 \leq t \leq T, |x - z|/(T - t) \leq M\},$$

where $M \geq K$. It will be seen that the values of a solution u in a cone of determinacy are completely determined by the values on its base, $B \cap D_{z, T}$. Define a horizontal section $B_{z, t, T}$ to be those points of $D_{z, T}$ with time component equal to t. Both the uniqueness of the solution and the continuous dependence upon the initial data follow from this crucial lemma.

LEMMA 2.1   Let H be of class C^2 and convex, and let u and v be (almost everywhere) solutions of the Hamilton–Jacobi equation for H in a common cone of determinacy for u and v, $D = D_{z,T}$. In D, let u and v have common Lipschitz constant P, and common constant of semiconcavity in x, $A/t + C$. Let $D_t = B_{z,t,T}$ be the horizontal plane section of D with altitude t. Define

$$\varphi = \max_{\mathscr{D}} \left| \sum_k \partial^2 H / \partial x^k \, \partial p^k \right|,$$

$$\lambda = \max_{\mathscr{D}} \sum_k \partial^2 H / \partial p^k \, \partial p^k,$$

where

$$\mathscr{D} = \{(t, x, p) : (t, x) \in D, |p| < P\}.$$

Then for every even integer m,

$$t^{-A\lambda} e^{-(\varphi + C\lambda)t} \int_{D_t} (u - v)^m \, dx$$

is a nonincreasing function of t in the interval $0 < t \le T$.

Proof.   Since u and v are solutions of the Hamilton–Jacobi equation, letting $w = u - v$ results in

$$w_t + G w_x = 0$$

almost everywhere in D, where G is the n-vector

$$G(t, x, u, v) = \int_0^1 H_p(t, x, v_x + \theta(u_x - v_x)) \, d\theta.$$

Thus $W = e^{-\varphi t} w^m$ satisfies

$$W_t + G W_x + \varphi W = 0 \tag{2.1}$$

almost everywhere in D.

Mollifying u and v, let u' and v' be approximations of u and v respectively, which are of class C^2 and which have the same Lipschitz constant P, and constant of semiconcavity in x, $A/t + C$, as u and v. Denote $\partial u'/\partial x^k$ simply by u_k' and so on, and define G' from u' and v' as G was defined from u and v. Equation (2.1) yields

$$W_t + G' W_x = (G' - G) W_x - \varphi W$$

or

$$W_t + (G'W)_x = (G' - G) W_x - \varphi W + W G_x', \tag{2.2}$$

where G_x' is a shorthand for

$$\sum_i \partial(G')^i/\partial x^i$$

It will now be shown that G_x' is of the form

$$G_x' = I + J$$

where $|I| \leq \varphi$ and $J \leq \lambda(A/t + C)$. Begin by defining

$$I = \int_0^1 \sum_k \partial^2 H(t, x, v_x' + \theta(u_x' - v_x'))/\partial x^k \, \partial p^k \, d\theta, \tag{2.3}$$

$$J = \int_0^1 \theta H_{pp} \, d\theta \times u_{xx}' + \int_0^1 (1 - \theta) H_{pp} \, d\theta \times v_{xx}', \tag{2.4}$$

where the arguments of H_{pp} in (2.4) are the same arguments as those in (2.3), and the multiplication is not matrix multiplication, but rather the sum of the products of all corresponding elements. That is, if $Q = (Q_{ij})$ and $R = (R_{ij})$ are matrices,

$$Q \times R = \sum_{ij} Q_{ij} R_{ij}.$$

Differentiating G' shows that $G_x' = I + J$ as desired. Also $|I| \leq \varphi$ by the definition of φ. To show that $J \leq \lambda(A/t + C)$, note that by assumption,

$$h^T u_{xx}' h \leq k = A/t + C$$

for all unit vectors h in $\mathbb{R}^n$.

Thus if δ represents the identity matrix (δ_{ij}),

$$h^T(u_{xx}' - k\delta)h \leq 0,$$

so the matrix $U = u_{xx}' - k\delta$ is nonpositive. By the convexity of H, the two integrals in (2.4) are nonnegative.

The two terms of (2.4) are similar, so we shall consider only the first. Let

$$Z = \int_0^1 \theta H_{pp} \, d\theta,$$

and rewrite the first term as

$$Z \times u_{xx}' = Z \times (u_{xx}' - k\delta) + kZ \times \delta$$
$$= Z \times U + \sum_i Z_{ii}$$
$$= \text{tr}(ZU) + k \sum_i Z_{ii},$$

where tr(ZU) denotes the trace of the matrix product ZU. Since the trace of the product of a symmetric nonpositive matrix and a symmetric nonnegative matrix is nonpositive,

$$Z \times u'_{xx} \le 0 + k \sum_i Z_{ii} \le k\lambda/2,$$

so that $J \le \lambda(A/t + C)$ as claimed.

We shall now integrate equation (2.2) over D_ϵ', the frustum of D between D_ϵ and D_t, where

$$0 < \epsilon < t \le T.$$

Let E_ϵ' denote the boundary of D_ϵ' excluding D_ϵ and D_t, that is, the sloping part of the boundary, and use dS and $(v_t, v) = (v_t, v_1, \ldots, v_n)$ to denote the element of area and the unit outward normal at a point of E_ϵ'. Then by the divergence theorem, we have

$$\int_{D_\tau} W \, dx - \int_{D_\epsilon} W \, dx + \int_{E_\epsilon'} W(v_t + G'v) \, dS$$

$$= \int_{D_\epsilon'} (G' - G)W_x \, dx \, dt + \int_{D_\epsilon'} (I - \varphi)W \, dx \, dt + \int_{D_\epsilon'} JW \, dx \, dt.$$

Now on E_ϵ', $v_t + G'v \ge 0$ and $I - \varphi \le 0$, so since $W \ge 0$,

$$\int_{D_\tau} W \, dx \le \int_{D_\epsilon} W \, dx + \lambda \int_{D_\epsilon'} (A/t + C)W \, dx \, dt$$

$$+ \int_{D_\epsilon'} (G' - G)W_x \, dx \, dt.$$

Now letting u' and v' be chosen so that $u' \to u$, $v' \to v$, $u_x' \to u_x$ and $v_x' \to v_x$ almost everywhere in D, we have

$$\int_{D_\tau} W \, dx \le \int_{D_\epsilon} W \, dx + \lambda \int_{D_\epsilon'} (A/t + C)W \, dx \, dt.$$

Let $\eta(t) = \int_{D_t} W \, dx$, to obtain

$$\eta(\tau) \le \eta(\epsilon) + \lambda \int_\epsilon^\tau (A/t + C)\eta(t) \, dt.$$

Thus η is majorized by

$$z(t) = (t/\epsilon)^{\lambda A} e^{\lambda C(t - \epsilon)} \eta(\epsilon)$$

for $\epsilon \le t \le T$ since $z(t)$ is a solution of

$$z(\tau) = \eta(\epsilon) + \lambda \int_\epsilon^\tau (A/t + C)z(t) \, dt.$$

But $\eta(t) \leq z(t)$ expands to

$$\int_{D_t} W \, dx \leq (t/\epsilon)^{\lambda A} e^{\lambda C(t-\epsilon)} \int_{D_\epsilon} W \, dx$$

or

$$e^{-\varphi t} \int_{D_t} w^m \, dx \leq (t/\epsilon)^{\lambda A} e^{\lambda C(t-\epsilon)} e^{-\varphi \epsilon} \int_{D_\epsilon} w^m \, dx,$$

which yields the desired result immediately. ∎

THEOREM 2.1 *Theorem on Continuous Dependence*   In the situation of lemma 2.1, for $0 \leq t \leq T$,

$$\max_{D_t} |u - v|$$

is a nonincreasing function of t.

Proof   By the lemma, for $0 < \epsilon < t \leq T$ and for even m,

$$\int_{D_t} (u - v)^m \, dx \leq (t/\epsilon)^{\lambda A} e^{(\varphi + \lambda C)(t-\epsilon)} \int_{D_\epsilon} (u - v)^m \, dx.$$

Raising both sides to the power $1/m$ and letting $m \to \infty$ yields

$$\max_{D_t} |u - v| \leq \max_{D_\epsilon} |u - v|$$

immediately. Since u and v are continuous at $t = 0$, this inequality is also valid at $\epsilon = 0$. ∎

COROLLARY   *Douglis' Uniqueness Theorem*   Suppose that H is convex and of class C^2 and suppose that two solutions (Lipschitzian in any S_T and semiconcave in x with constant $A/t + C$ agree on the base of a common cone of determinacy. Then the solutions agree throughout the cone. If the solutions coincide on the entire hyperplane $t = 0$, then they are the same for all (t, x) with $t \geq 0$. ∎

3. UNIQUENESS FOR MORE GENERAL BOUNDARIES

The result of Douglis in section 2 suggests that the solution, at least in the semiconcave case, is determined uniquely in any cone of determinacy which completely covers a portion of the boundary. That is, if u and v are solutions with K a common bound on the characteristic slopes, and any Lipschitzian path $\alpha: [0, t] \to \mathbb{R}^n$ ending at (t, x) with $\|\alpha'\|_\infty \leq K$ has a point

$(s, y) = (s, \alpha(s))$ with $u(s, y) = v(s, y)$, then we would expect that $u(t, x) = v(t, x)$. The author conjectures that this is true in quite general cases, but has not yet found time to investigate this question. However, recent results of Feltus [101] show that this is true for certain mixed problems.

For the remainder of this section we shall follow the development of Feltus. The boundary will consist of two pieces,

$$B = B^0 \cup B',$$

where

$$B^0 = \{0\} \times X,$$

$$B' = [0, \infty) \times \partial X,$$

with X a bounded domain in $\mathbb{R}^n$ with smooth boundary ∂X. As before S_T will denote that part of D with $0 \leq t \leq T$, but D will be only $(0, \infty) \times X$. The Hamiltonian will be $H = H(p)$ only and of class C^2. The data $f(t, x)$ will be equal to $\varphi(x)$ on B^0 and equal to $\gamma(t, x)$ on B'. It will also be assumed that H is strictly convex. Thus for nonzero $h \in \mathbb{R}^n$,

$$h^T H_{pp}(p)h \geq a(p)|h|^2 > 0.$$

Now let us fix $T > 0$ and define, for $K > 0$, $\mathscr{L}(K)$ to be the class of all measurable functions $S_T \to \mathbb{R}$ with both supremum norm and Lipschitz constant bounded above by K. A solution of the problem (H, f) (in the sense of Feltus) is a real function $u \in \bigcup_K \mathscr{L}(K)$ which satisfies the Hamilton–Jacobi equation almost everywhere in S_T, is equal to f on B (any such f may be extended continuously to B), and is uniformly semiconcave in the sense that, for some fixed k,

$$u(t, x + h) + u(t, x - h) - 2u(t, x) \leq k|h|^2$$

for all sufficiently small h. Note that the phrase "for all sufficiently small h" modifies our previous usage and makes it easier to establish the condition, as we shall see in the exercises. For every generalized solution u, define M_u to be the maximum of the supremum of $|H_p(p)|$ and that of $\sum_j \partial^2 H(p)/\partial p_j \, \partial p_j$ for $|p| \leq K$, where K is the least K with $u \in \mathscr{L}(K)$.

THEOREM 3.1   *Feltus' Uniqueness Theorem*   There is at most one solution (in the sense of Feltus) to the above problem.

Proof   Assume that u and v are both solutions. As before, the difference, $w = u - v$ satisfies

$$w_t + Gw_x = 0 \qquad \text{a.e.,} \tag{3.1}$$

where

$$G = \int_0^1 H_p(\theta u_x + (1 - \theta)v_x) \, d\theta. \tag{3.2}$$

Letting $W = |w|$, W is also Lipschitzian and satisfies (3.1) almost everywhere in S_T.

Now let u' and v' be smooth (C^2) approximations of u and v respectively, with the same bounds, Lipschitz constants, and common constant of semi-concavity k. Denote $\partial^2 u'/\partial x^i \, \partial x^j$ by u'_{ij} and so on, as before, and define G' from u' and v' as in (3.2). We now obtain

$$W_t + G'W_x = (G' - G)W_x,$$

so that

$$W_t + G'W_x + G_x'W = (G' - G)W_x + G_x'W$$

and

$$W_t + (G'W)_x = (G' - G)W_x + WG_x', \tag{3.3}$$

where, as before, G_x' denotes $\sum_i \partial(G')^i/\partial x^i$.

Integrating (3.3) over S_T and using the divergence theorem and letting D_t be as in Douglis' proof yields us

$$\int_{D_t} W \, dx = \int_{S_T} (G' - G)W_x \, dx \, dt + \int_{S_T} WG_x' \, dx \, dt. \tag{3.4}$$

As in Douglis' proof, we have

$$G_x' \le k(M_u + M_v) = kM.$$

So (3.4) leads to

$$\int_{D_t} W \, dx \le \int_{S_T} (G' - G)W_x \, dx \, dt + kM \int_{S_T} W \, dx \, dt. \tag{3.5}$$

Now the $\int (G' - G)W_x$ is bounded above in absolute value by $\|G' - G\|_1 \cdot \|W_x\|_\infty$, where $\|G' - G\|_1$, denotes the L_1 norm of $|G' - G|$ and $\|W_x\|_\infty$ denotes the L_∞ norm of $|W_x|$. Letting $u' \to u$ and $v' \to v$ as in Douglis' proof, we have

$$\int_{D_t} W \, dx \le kM \int_{S_T} W \, dx \, dt. \tag{3.6}$$

Letting $z(t) = \int_{D_t} W \, dx \, dt$, (3.6) becomes

$$z'(\tau) \le kMz(\tau), \qquad 0 < \tau < T,$$

or z is dominated by $z(0)\exp(kM/\tau)$. But $z(0) = 0$, so $W = 0$ almost everywhere in S_T, as was to be shown. For more details, see Feltus [101, pp. 12–21]. ∎

Feltus also obtained a similar uniqueness theorem for a mixed problem where X is unbounded. Namely, X is an "octant"

$$X = \{(x^1, x^2, \ldots, x^n) \in \mathbb{R}^n : x^i > 0, 1 \le i \le n\}.$$

The proof is somewhat similar to the above proof and may be found in [101, pp. 22–26]. Feltus then established that, at least for a special one-dimensional mixed problem, the variational solution possesses the necessary properties for the uniqueness theorem, that is, is the "right" solution.

The methods of section 1 of this chapter may be used to establish the necessary semiconcavity properties. For example, if φ may be extended to all of $\{0\} \times \mathbb{R}^n$ so that theorem 1.1 (7) applies, and B' is a finite number of hyperplanes determining an "octant" so that γ may be extended to all of $\mathbb{R} \times \partial X$ with theorem 1.2 (9) applying to each hyperplane, the variational solution, as a minimum of a finite number of uniformly semiconcave functions, is itself uniformly semiconcave. That is, the variational problem for each hyperplane of the boundary is solved separately, and the solution shown to be uniformly semiconcave in x where desired. The final variational solution is the minimum of these partial variational solutions.

Exercises

3.1   In the one-dimensional case, where $B = \{0\} \times [0, \infty)$ and $B' = [0, \infty) \times \{0\}$, the necessary extensions are easy, so that the concluding remarks of this section are simple to rigorize. Summarize them in a theorem.

3.2   Give theorems for n-dimensional problems which establish the uniform semiconcavity of the variational solution in special cases.

3.3   Prove or disprove the author's conjecture from the beginning of this section. Send your solution to the author, and publish it unless you want it plagiarized.

4. MAXIMALITY OF THE VARIATIONAL SOLUTION

Even granting the author's conjecture from section 3, it is not immediately apparent whether or not global solutions might be uniquely determined in the classical boundary value problem. They are not, as the following example shows.

Example 4.1   Consider the one-dimensional boundary value problem,

$$u_t + u_x^2/4 = 0, \qquad t > 0, \quad x \in \mathbb{R}$$

$$u(t, 0) = 0, \qquad t \geq 0.$$

Here $B = \mathbb{R} \times \{0\}$, the t-axis, $H(p) = p^2/4$, and $L(q) = q^2$, $f(t, 0) \equiv 0$. The variational solution is

$$u(t, x) = x^2/t, \qquad t > 0.$$

This is a C^∞ solution in $t > 0$, and is semiconcave in x with constant $2/t$.

But suppose we now add to B the point $(-1, 0)$ and set $f(-1, 0) = a \geq 0$. The variational solution now becomes

$$v_a(t, x) = a + x^2/(t + 1), \qquad t > -1.$$

This solution is of class C^∞ in $t > -1$, and is semiconcave in x with constant $2/(t + 1)$.

Now $u_a = \min(u, v_a)$ is a solution of the original problem for $t > 0$, semiconcave in x with constant $2/t$. Specifically,

$$u_a(t, x) = \begin{cases} x^2/t, & x^2 \leq a(t^2 + t), \quad t > 0, \\ a + x^2/(t + 1), & x^2 > a(t^2 + t), \quad t > 0. \end{cases}$$

So $u_a \neq u_c$ for $a \neq c$. Define $u_\infty = u$. Then $\{u_a : 0 \leq a \leq \infty\}$ is an uncountable family of global solutions of the original problem, all of which are of class C^∞ except where $x^2 = a(t^2 + t)$, and all of which are semiconcave in x with constant $2/t$. Note that $u \equiv 0$ is another, completely unrelated, solution with these same properties.   ∎

In example 4.1, the problem seems to be the introduction of data from points below D. One might think that allowing B to extend to $t = -\infty$ (that is, $b = -\infty$) might give uniqueness for the boundary value problem. Consideration of the two solutions

$$u_1 = x - t,$$

$$u_2 = -x - t,$$

to the problem

$$u_t + u_x^2 = 0, \qquad (t, x) \in \mathbb{R}^2,$$

$$u(t, 0) = -t, \qquad t \in \mathbb{R},$$

shows that this is not the case either.

Thus uniqueness of solutions for general boundaries is out of the question. However it does happen that the variational solution is the greatest possible solution. That is, we shall show that, if u is any global solution of the problem (H, f), and if v is the variational solution, then for each point (t, x) of D,

$$v(t, x) \geq u(t, x).$$

we will do this by letting $\alpha : [s, t] \to \mathbb{R}^n$ be any Lipschitzian curve and showing that

$$u(t, x) - u(s, y) \leq \int_\alpha L, \tag{4.1}$$

where $x = \alpha(t)$, $y = \alpha(s)$, and $u(s, y)$ is defined to be $f(s, y)$. Thus

$$u(t, x) \leq f(s, y) + \int_\alpha L,$$

and since this holds for any α, it holds for the infimum over all α, that is,

$$u(t, x) \leq v(t, x).$$

Suppose temporarily that the entire graph of α is contained in D, rather than having $(s, y) \in B$, and consider the heuristic argument of section 2, chapter II. Since u is locally Lipschitzian, α Lipschitzian, and the graph of α is compact, it follows that $U(\tau) = u(\tau, \alpha(\tau))$ is Lipschitzian, hence absolutely continuous. This yields the validity of (II.2.2). Now (II.2.3) and (II.2.4) hold almost everywhere in D, although perhaps nowhere along α. To eliminate this problem, let us approximate α by other curves which miss the bad points of D. Toward this end, for $\tau \in [s, t]$, $\epsilon > 0$, and $z \in \mathscr{B}$, the open unit ball in $\mathbb{R}^n$, define

$$\beta(\tau, z) = \alpha(\tau) + \delta(\tau, z),$$

$$\delta(\tau, z) = 2\epsilon \min(t - \tau, \tau - s)(t - s)^{-1}z.$$

Let C denote the space swept out by the curves defined by β. That is,

$$C = \{(\tau, w): s < \tau < t, |w - \alpha(\tau)| < 2\epsilon \min(t - \tau, \tau - s)/(t - s)\}.$$

LEMMA 4.1   Let N be a set of Lebesgue measure zero in $\mathbb{R} \times \mathbb{R}^n$, $\epsilon > 0$, and β as defined above. Then for almost all $z \in \mathscr{B}$, the set

$$\mathscr{S} = \{\tau \in (s, t) : (\tau, \beta(\tau, z)) \in N\}$$

has one-dimensional Lebesgue measure zero.

Proof   Let I be the indicator function of N, C as above, let μ denote $(n + 1)$-dimensional Lebesgue measure, and let $\mathscr{B}'$ denote the set $2\epsilon \min(t - \tau, \tau - s)/(t - s)\mathscr{B}$. Then the change of variables formula,

Fubini's theorem, and the translation invariance of Lebesque measure imply the following sequence:

$$0 = \mu(N \cap C)$$

$$= \int_C I \, d\mu$$

$$= \int_s^t \int_{\mathscr{B}' + \alpha(\tau)} I(\tau, z) \, dz \, d\tau$$

$$= \int_s^t \int_{\mathscr{B}'} I(\tau, \alpha(\tau) + z) \, dz \, d\tau$$

$$= \int_s^t \int_{\mathscr{B}} I(\tau, \alpha(\tau) + \delta(\tau, z)) |J| \, dz \, d\tau$$

$$= \int_{\mathscr{B}} \int_s^t I(\tau, \beta) |J| \, d\tau \, dz,$$

where J is the nonsingular Jacobian determinant of the mapping $(\tau, z) \mapsto (\tau, \delta(\tau, z))$. This mapping is continuously differentiable except at $\tau = (s + t)/2$, so that the change of variables formula applies. Since the last integral is zero, for almost all $z \in \mathscr{B}$, the inside integral vanishes. Since J is not zero for all τ in $(s, t) \setminus \{(s + t)/2\}$, the vanishing of the inside integral implies that $I(\tau, \beta)$ vanishes for almost all τ. That is, for almost all τ, $(\tau, \beta(\tau, z))$ is not in N, as claimed. ∎

LEMMA 4.2   Let L be continuous, α a Lipschitzian curve in D joining (t, x) to (s, y) with $s < t$. Then if u solves (H, f),

$$u(t, x) - u(s, y) \leq \int_\alpha L.$$

Proof   Let N be the null set consisting of points of D where II.2.3 or II.2.4 does not hold. Choose ϵ so that the set C as previously defined is contained in D. Note that the arguments τ, $\beta(\tau, z)$, $\beta'(\tau, z)$ are all bounded on C, so that L is uniformly continuous for these arguments. For each $\epsilon' \in (0, \epsilon)$, choose a curve $\beta : [s, t] \to \mathbb{R}^n$ such that $(\tau, \beta(\tau))$ is not in N for almost all τ, and $\| \alpha' - \beta' \|_\infty < \epsilon'$, by lemma 4.1. Then

$$u(t, x) - u(s, y) \leq \int_\beta L \to \int_\alpha L \qquad \text{as} \quad \epsilon' \to 0. \qquad ∎$$

Now if (s, y) is a boundary point, but the rest of the graph of α is in D, any approach $(t_k, x_k) \to (s, y)$ along the graph of α is deterministic since α is Lipschitzian, thus $u(t_k, x_k) \to u(s, y) = f(s, y)$ and along α,

$$\int_{t_k}^{t} L \to \int_{s}^{t} L = \int_{\alpha} L,$$

thus

$$u(t, x) - f(s, y) \le \int_{\alpha} L.$$

Finally, if $(t, x) \in D$, let σ be the maximum time value τ for which $(\tau, \alpha(\tau)) \in B$. If the compatibility condition holds, then

$$f(\sigma, \alpha(\sigma)) - f(s, y) \le \int_{s}^{\sigma} L,$$

and by the above,

$$u(t, x) - f(\sigma, \alpha(\sigma)) \le \int_{\sigma}^{t} L.$$

Combining these,

$$u(t, x) - f(s, y) \le \int_{\alpha} L,$$

as was claimed. Thus we have the following theorem.

THEOREM 4.1   *Maximality of the Variational Solution*   Let (H, f) be a Hamilton–Jacobi boundary value problem such that L is at least continuous and the compatibility condition holds. Then if u is any solution, $(t, x) \in D$, $(s, y) \in B$ with $s < t$, and α is any Lipschitzian path joining (t, x) to (s, y), it follows that

$$u(t, x) \le f(s, y) + \int_{\alpha} L. \quad \blacksquare$$

COROLLARY   In the situations of the first and second existence theorems, the variational solution is greater than or equal to any other solution at each point of D. $\blacksquare$

IV Applications and Numerical Methods

GENERAL COMMENTS

To claim any semblance of completeness for either applications or numerical methods for the Hamilton–Jacobi equation is patently absurd. Certainly the pertinent numerical methods could fill a volume of this size by themselves, as could any of several of the applications. Mechanics, for instance, could fill several volumes. Thus we will make little attempt to develop the physical applications, except to summarize a few of them in section 1. Global techniques have not been applied in any systematic manner to these physical applications, so may provide a fertile research area for applied mathematicians in the near future.

Sections 2–4 deal primarily with mathematical applications, with no particular physical application in mind, and do require global solutions. Sections 5–7 summarize a few numerical methods, primarily difference methods, and "artificial viscosity" or second-order parabolic methods. This concluding chapter will be brief and proofs will be omitted. Its primary goal is to point the reader toward the extensive literature in these areas. Even in this, we shall fall far short of completeness because of the massive volume of available references. Should the author claim to have perused even all of the major references, the typical computer science type of argument, assuming one book per hour or some such thing, would certainly show that 10^6 years or thereabouts were needed for the reading. However, an attempt has been made to list a few of the classic texts, or a few well presented modern expositions, for each topic mentioned.

1. SOME PHYSICAL APPLICATIONS

Probably the most well known use of the calculus of variations is the Hamiltonian, or Lagrangian, treatment of mechanics. In the simplest case, where the "path" desired is an actual trajectory in a conservative force field in Euclidean 3-space, the Lagrangian is the "action"

$$L = T - V,$$

where T represents the specific kinetic energy of the system and V represents the potential. The extremals of the corresponding variational problem furnish the desired paths. If the Hamiltonian H is defined as the Legendre conjugate of L, these extremals are the characteristics of the Hamilton–Jacobi equation. Thus the Hamilton–Jacobi equation comes into the problem in a natural manner. As we have noted in chapter I, a complete integral leads immediately to the extremals via a contact transformation.

A slight generalization of this procedure allows use of "generalized coordinates," rather than the Euclidean coordinates of the parts of the system. For details, see Goldstein [122]. Within the realm of classical mechanics, the method may be generalized to nonconservative forces by replacing the " $-V$ " term by a "virtual work" term. Thus almost all of classical mechanics may be treated as a variational problem or as a Hamilton–Jacobi equation, including problems in elasticity, elementary particles, and electromagnetic fields.

Relativistic mechanics may also be given a Lagrangian formulation. The problem is to find a suitable Lagrangian L. This L is no longer $T - V$ but is still a function whose partial derivative with respect to time is the momentum. For example, Goldstein [122, p. 206] gives a relativistic Lagrangian for a single particle in a conservative force as

$$L = -mc^2\sqrt{1 - \beta^2} - V,$$

where m is the mass, c the velocity of light, β the normalized velocity ($c = 1$), and V represents the potential.

Even most of quantum mechanics has been given the Lagrangian formulation. The connection comes by choosing as a Hamiltonian, the total energy, and deriving geometrical optics and wave mechanics from this direction. Again, see Goldstein [122] for details. For further reading, see Hermann [136], or any other text on quantum mechanics.

As previously stated, we shall make no attempt to develop these applications here, but merely point them out to indicate part of the range of applicability of the Hamilton–Jacobi equation. The reader is also referred to Saletan and Cromer [234] for more mechanics, Fitzpatrick [105] or Pollard

[217] for applications in celestial mechanics, Petrashen and Trifonov [209] for more quantum mechanics, Washizu [263] for variational methods in elasticity and plasticity, and Kazuo Kitahara's article in Nicolis and Lefever [201, pp. 85–111] for a short treatment of the Hamilton–Jacobi approach to fluctuation phenomena.

2. APPLICATIONS IN THE CALCULUS OF VARIATIONS AND CONTROL THEORY

The relation of the Hamilton–Jacobi equation to the calculus of variations has been considered extensively in this work so far. However, the emphasis was on solving the PDE, at least in theory, by solving the variational problem. Here we would like to note the reverse problem. That is, confronted with the problem of minimizing

$$J(\alpha) = f(y) + \int_\alpha L$$

over all Lipschitzian curves $\alpha: [0, t] \to \mathbb{R}^n$, where $\alpha(0) = y$, one might solve the associated Hamilton–Jacobi equation first. If the solution u, be it analytic or numeric, is semiconcave in x with a constant $A/t + C$ and the hypotheses of the first or second existence theorem and Douglis' uniqueness theorem hold, then u is also the required variational minimum. The extremals may then be found by solving the ODE

$$\alpha'(\tau) = H_p(\tau, \alpha(\tau), u_x(\tau, \alpha(\tau))),$$

or from the Euler equations.

Now control problems are, at first glance, considerably more complicated than simple variational problems. However, quite general control problems may be put in a variational format, thus allowing the Hamilton–Jacobi equation to solve control problems as well as variational problems. To demonstrate this point, we present the following quite general control problem from Saaty and Bram [231, pp. 292–293], which they attribute to Berkovitz [34] and [35]. The problem is to find vector functions

$$x(t) \in \mathbb{R}^n, \qquad y(t) \in \mathbb{R}^m,$$

which minimize the functional

$$J(x, y) = \int_{t_0}^{t_1} F(t, x, y') \, dt,$$

where $s \in \mathscr{D} \subseteq \mathbb{R}^p$ is a "parameter" which determines the endpoints for t and x, subject to the restrictions:

(1) $x'(t) = f(t, x(t), y'(t))$.
(2) $G(t, x, y') \geq 0$,

where $G \in \mathbb{R}^r, r \leq m$.

(3) $\theta(t, x) \geq 0$.
(4) $x(t_0) = x_0, \qquad y(t_0) = y_0 = 0$,

$$t_1 = t_1(s), \qquad x_1 = x(t_1) = x_1(s), \qquad y_1 = y(t_1).$$

Other end conditions than **(4)** above may be added, and $r > m$ can also be considered as long as certain rank conditions are met. Reduce **(2)** to equalities by setting

$$G^i = (z_i')^2, \qquad 1 \leq i \leq r,$$

$$z(t_0) = 0.$$

Denote this simply by $G = (z')^2$. Now define

$$\gamma(t, x, y) = \begin{cases} \eta^4 - \theta(t, x), & \eta \geq 0, \\ -\theta(t, x), & \eta \leq 0, \end{cases}$$

and replace **(3)** by

$$\theta_t + \theta_x x' - \gamma_\eta \eta' = 0,$$

and add the additional end conditions

$$\gamma(t_0, x_0, \eta_0) = 0, \qquad \gamma(t_1, x_1, \eta_1) = 0,$$

$$\eta(t_0) = \eta_0, \qquad \eta(t_1) = \eta_1.$$

Now we may form the Lagrangian

$$L(t, x, y, z, x', y', z', \lambda_0, \lambda(t), \mu(t), v(t)) = \lambda_0 F + \lambda(t)(f - x') + \mu(t)(G - (z')^2)$$
$$+ v(t)(\theta_t + \theta_x x' - \gamma_\eta \eta'),$$

where $\lambda_0 \geq 0$ and $\lambda(t) \in \mathbb{R}^n$, $\mu(t) \in \mathbb{R}^r$, $v(t) \in \mathbb{R}$, and $(\lambda_0, \lambda, \mu, v)$ is continuous on (t_0, t_1) and vanishes only at isolated points (corners). Solving the given control problem becomes "equivalent" to solving the variational problem with Lagrangian L. Saaty and Bram apparently allow $q > 1$ of the constraints in **(3)**, and allow a slightly more general functional J, indicating that quite general control problems may be reduced to variational problems.

Since many dynamic programming problems are actually control problems, or are closely related, the Hamilton–Jacobi equation also finds applications in dynamic programming. See Angel and Bellman [12] and

Bellman [24] for an introduction to these applications, as well as further information on control theory and the calculus of variations.

Another area closely related to control theory, where the Hamilton–Jacobi equation finds applications is the theory of differential games. For the connections, the reader is referred to Elliott [94], and Elliott and Kalton [96, 98, 99].

3. MINIMIZATION OF A FUNCTION

In the case of the space–time independent Hamiltonian, $H(u_x)$, the desired solution is simply a minimum of a function of several variables. For instance, for the Cauchy problem where $B = \{0\} \times \mathbb{R}^n$, a solution is obtained by minimizing, over all $y \in \mathbb{R}^n$, the function

$$g(y) = f(y) + tL((x - y)/t). \tag{3.1}$$

In some trivial cases, g may be minimized algebraically.

Example 3.1   Consider the problem in n dimensions

$$u_t + |u_x|^2/4 = 0, \qquad t > 0,$$

$$u(0, x) = a|x|^2 + bx + c,$$

where $a > 0$, $b \in \mathbb{R}^n$, and c is real. The function to be minimized is

$$g(y) = a|y|^2 + by + c + |x - y|^2/t.$$

Completing the square, this is

$$(a + 1/t)^{-1}g(y) = |y + (tb - 2x)/2(at + 1)|^2 + (ct + |x|^2)/(at + 1)^2$$
$$- |tb - 2x|^2/4(at + 1)^2.$$

This function is clearly minimized when

$$2(at + 1)y = 2x - tb,$$

yielding

$$u(t, x) = c + ad|x|^2 + d(bx - t|b|^2/4),$$

where $d = 1/(at + 1)$.   ∎

Of course the most common method of minimizing $g(y)$ is to set its gradient equal to zero.

Example 3.2   To solve the problem of example 3.1 by calculus methods, set

$$g_y(y) = 2ay + b + 2(y - x)/t = 0,$$

obtaining

$$2(a + 1/t)y + b - 2x/t = 0,$$

and, of course, the same solution as before.   ∎

In case B is a smooth manifold, but not necessarily $[t = 0]$, the method of Lagrange multipliers is generally resorted to for an analytic solution. See Saaty and Bram [231] or almost any textbook on several variable calculus or applied mathematics. However, in any case, analytic determination of the desired minimum is impossible in any case of nontrivial complexity. Thus numerical methods are usually sought.

For finding such a minimum, the gradient method, or method of steepest descent is often used. This method consists of starting with any value of y, say y_0, and using the fact that the direction of fastest increase of $g(y)$ is the direction of the gradient $g_y(y)$. Thus y_1 is a point

$$y_1 = y_0 - \lambda g_y(y_0),$$

where λ is a positive constant which may be chosen in several ways. The optimal method is to choose λ to minimize $g(y_1)$. If this is not feasible, on may try various values for λ by some iterative scheme. Or, letting λ become "infinitesimal," one may use the differential equation

$$y'(\lambda) = -g_y(y).$$

Given proper hypotheses on g, $y(\lambda)$ may approach the optimal y as $\lambda \to \infty$. For more details on the gradient method, see Fleming [107] or a text on numerical analysis. Generally speaking, choosing the best λ, and iterating leads to a sequence

$$y_0, y_1, y_2, \ldots,$$

which converges to the optimal y if g is convex, and sometimes converges even if y is not convex if y_0 is chosen correctly.

Other numerical methods attack the equation

$$g_y(y) = 0 \tag{3.2}$$

to find a root, which is hopefully the optimum value of y. Picard iteration, with several improvements, is a method which searches for a fixed point of

$$G(y) = g_y(y) + y.$$

See Pizer [214], Williamson [265, 266], or a numerical analysis text. The method is simply to choose a y_0 as before and form

$$y_1 = G(y_0),$$

$$y_2 = G(y_1) = G^2(y_0),$$

$$\vdots$$

$$y_m = G^m(y_0), \quad \text{etc.}$$

If G is Lipschitzian with constant less than 1, and its range is included in its domain (or y_0 is suitable restricted), $y_m \to y_\infty$, the optimal y, as $m \to \infty$. This is known as the contraction mapping principle. See Knill [157] for generalizations.

Other root finding methods in the one-dimensional case are plentiful. They include "bisection," "Newton's method," "regula falsi," and many generalizations. See Scheid [236], Henrici [135], or Pizer [214]. Most of these have multidimensional generalizations.

Another possibility is using the gradient method on

$$e(y) = |g_y(y)|^2,$$

instead of on $g(y)$ directly. The advantages of this approach are that e may be convex where g is not, and that λ may be chosen by a method similar to Newton's method since it is known that the desired minimum is zero. That is,

$$y_1 = y_0 - \lambda g_y(y_0),$$

where

$$\lambda = g(y_0)/|g_y(y_0)|^2.$$

This is generally simpler than choosing λ to minimize $g(y_1)$.

We see, then, that our representation (3.1) of the solution leads to a host of numerical and analytic methods. However most of these methods are complicated and work only in special cases. Thus we come to the point of this section. We offer yet another method, based upon our global solutions. Just as Jacobi reversed the method of characteristics, using a PDE to solve an ODE problem, we use a PDE problem to solve a minimization problem.

Suppose then, that we wish to minimize a real function, $g: \mathbb{R}^n \to \mathbb{R}$. Suppose that g is convex and of class C^2. Set

$$L(y) = g(-y),$$

and let H be the convex dual of L. Now consider the problem

$$u_t + H(u_x) = 0,$$

$$u(0, x) = 0.$$

Douglis' uniqueness theorem applies here, so if a solution can be found, either numerically or analytically, which is semiconcave in x with a constant of the form $A/t + C$ with A and C positive, it must be the variational solution

$$u(t, x) = \min_y\{tL((x - y)/t)\}.$$

Setting $t = 1$ and $x = 0$ (the global solution extends to $t = 1$ by definition), we have

$$u(1, 0) = \min_y g(y).$$

Thus the solution of our PDE problem gives the desired minimum. If the optimal value of y is desired, any root finding method may be applied to

$$h(x) = g(x) - \min_y g(y) = g(x) - u(1, 0).$$

Sections 5, 6, and 7 of this chapter will summarize a few numerical methods of solving the Hamilton–Jacobi equation globally, which thus become minimization methods. While the author readily admits that this may be driving a carpet tack with a pile driver, he still thinks it is cute as a pickle seed.

Note also that our entire variational theory may be applied to functions H (or L) which are so-called "proper convex functions." That is, are defined not on all of $\mathbb{R}^n$, but on a convex subset $\mathscr{C}$, are convex, and tend to ∞ at the boundary of $\mathscr{C}$. Defining H (or L) to be ∞ off $\mathscr{C}$, nearly the same theorems obtain, although the proofs must be modified slightly. Thus this method of minimization is even more generally applicable than originally claimed.

4. QUASILINEAR EQUATIONS

Many quasilinear equations occurring in the applications may be treated directly as Hamilton–Jacobi equations. That is, they are actually linear, and are of the form

$$a(t, x)u_t + b(t, x)u_x = c(t, x),$$

which may be written as

$$u_t + (b/a)u_x - c/a = 0,$$

as long as a does not vanish. However, quasilinear equations have their own global theory and are usually easier to handle directly. In fact, our theory, requiring that $H(t, x, p)$ be convex in p with a growth condition, does not even apply to the linear case.

The transformation $p = u_x$ in the case that $n = 1$, establishes an "equivalence" between a quasilinear equation and a Hamilton–Jacobi equation, as outlined in chapter I, section 9. However that equivalence depends upon u being of class C^2, so that the mixed partial derivatives can be equated, hence is local in nature. The question naturally arises as to just how far the equivalence does extend in the case of global solutions. We shall see that the equivalence does, in a sense to be made explicit soon, carry over.

For simplicity, let us consider only the equation

$$v_t + \partial[H(v)]/\partial x = 0, \tag{4.1}$$

subject to

$$v(0, x) = g(x). \tag{4.2}$$

During the early work on global solutions of quasilinear equations, several different definitions of a weak solution of (4.1)–(4.2) were used. E. D. Conway, in notes from a 1968 analysis seminar at Tulane University, gives the following three definitions.

DEFINITION I   A weak solution of (4.1)–(4.2) in $D = [0, \infty) \times \mathbb{R}$ is a function $v \in \mathscr{L}_\infty(D)$ such that

$$\iint_D \{\varphi_t v + \varphi_x H\} \, dt \, dx + \int_{\mathbb{R}} \varphi(0, x) g(x) \, dx = 0$$

for every $\varphi \in C_0^{1}(D)$.

DEFINITION II   A weak solution of (4.1)–(4.2) in D is a function $v \in \mathscr{L}_\infty(D)$ such that

(a) $\quad \iint_D \{\varphi_t v + \varphi_x H\} \, dt \, dx = 0, \qquad \forall \varphi \in C_0^{1}(\text{int } D)$

and

(b) $\quad \displaystyle\int_{\mathbb{R}} \omega(x) v(t, x) \, dx \xrightarrow[t \to 0]{} \int_{\mathbb{R}} \omega(x) g(x) \, dx, \qquad \forall \omega \in C_0^{1}(\mathbb{R}).$

DEFINITION III   A weak solution of (4.1)–(4.2) in D is a function $v \in \mathscr{L}_\infty(D)$ such that

(a)   there is a function u, Lipschitzian in D, such that $v = u_x$ and $u_t + H(u_x) = 0$ a.e. in D.

(b) $\displaystyle\lim u(t, x) = \int_0^y g \qquad$ as $\qquad (t, x) \to (0, y).$

The motivation for the first two definitions may be seen by assuming a smooth solution and integrating by parts or by a brief study of the theory of distributions. Conway then shows that, for Hamiltonians H of class C^2, definitions I, II, and III are equivalent. The proof is a good exercise in analysis, using fairly straightforward arguments with mollifying functions, Ascoli's theorem, and weak convergence.

If g is bounded, so that $f(y) = \int_0^y g$ is Lipschitzian, our global solution of the problem (H, f) is Lipschitzian. (Also in many other cases. See chapter III.) Thus definition III above is our definition and our existence and uniqueness theorems may be applied to definition I- or II-type solutions of the quasilinear equation. Actually our definition, using $f(y)$ as $\int_0^y g$ may be used directly as a definition of a weak solution of (4.1)–(4.2). Thus our Hamilton–Jacobi theory may be used to study quasilinear equations.

In the case $n > 1$, our global solution of (H, f), where $B = \{0\} \times \mathbb{R}^n$ may be used as a definition of a weak solution of the quasilinear system

$$p_t + [H(t, x, p)]_x = 0,$$

$$p(0, x) = f_x(x),$$

where $p = u_x$ as before. However, this system is so specialized that the usefulness of this approach is questionable.

For more on global solutions of systems of equations, see Kuznetsov and Rozhdestvenskii [167] or the various papers of P. D. Lax which are listed in the references at the end of this work.

5. NUMERICAL METHODS FOR THE CALCULUS OF VARIATIONS

These last three sections of this work are intended to direct the reader to sources detailing numerical methods which may be used to solve the Hamilton–Jacobi equation. This section considers briefly a few methods which are applied directly to the variational problem.

Most of the numerical techniques for variational problems utilize the direct methods of the calculus of variations. That is, a minimizing sequence is chosen, shown to converge to a limiting curve, and lower semicontinuity is involved to ensure that the limiting curve does indeed furnish a minimum. The difference in the various methods lies primarily in the choice of the minimizing sequence. The best known of these procedures are the Ritz method and the Galerki method. We will outline the Ritz method for a one-dimensional problem below. For more details, see Akhiezer [5, pp. 143–148] and the references included there.

Suppose it is desired to minimize the functional

$$J(\alpha) = \int_s^t L(\tau, \alpha, \alpha')\, d\tau,$$

subject to the constraints that $\alpha(s) = y$ and $\alpha(t) = x$, and assume that sufficient hypotheses have been placed upon L so that minimizing sequences do converge to extremals. See the second existence theorem, or Akhiezer, for such hypotheses.

The Ritz method depends upon obtaining a sequence of functions of class C^1 $\{\varphi_k\}_{k=0}^\infty$, satisfying

(1) $\varphi_0(s) = y, \qquad \varphi_0(t) = x.$

(2) $\varphi_k(s) = \varphi_k(t) = 0, \qquad k > 0.$

(3)   The family $\{\varphi_k'\}_{k=1}^\infty$ is linearly independent.

(4)   Every continuous function $f:[s, t] \to \mathbb{R}$ is a uniform limit of "polynomials"

$$f_n(\tau) = \sum_{i=1}^n C_i \varphi_i'(\tau).$$

For φ_0, the line segment joining (t, x) and (s, y) is a logical place to start:

$$\varphi_0(\tau) = y + (x - y)(\tau - s)/(t - s).$$

For the remainder of the φ's, Akhiezer gives two possible choices as examples:

$$\varphi_k(\tau) = (\tau - s)^k(\tau - t), \qquad k > 0,$$

or

$$\varphi_k(\tau) = \sin(k\pi(\tau - s)(t - s)^{-1}), \qquad k > 0.$$

Given the sequence of φ's, the Ritz method consists of forming linear combinations,

$$\alpha_k(\tau) = \varphi_0(\tau) + \sum_{i=1}^k C_i \varphi_i(\tau),$$

and minimizing

$$G(C_1, C_2, \ldots, C_k) = J(\alpha_k).$$

This, of course, is simply a problem of minimization of a function of k variables. The optimal $C_1, C_2, \ldots, C_k$ yield a curve α_k and the sequence $\alpha_1, \alpha_2, \ldots$ becomes the desired minimizing sequence. For a proof of convergence to an extremal in a class of absolutely continuous curves, see Akhiezer.

Another method for reducing the problem to a problem of minimizing a function of k variables is even simpler. Simply choose $x_1, x_2, \ldots, x_k$, break the interval $[s, t]$ into $k + 1$ equal intervals by a partition

$$s = t_0 < t_1 < \cdots < t_k < t_{k+1} = t.$$

Let $x_0 = y$ and $x_{k+1} = x$. Then join (t_i, x_i) to (t_{i+1}, x_{i+1}) by straight line segments, $0 \le i \le k$. The resulting curve α_k is a broken line segment joining (t, x) to (s, y), and

$$G(x_1, \ldots, x_k) = J(\alpha_k)$$

may be minimized as in the Ritz method.

The final method to be mentioned here may be looked upon as a variant of the above method. The author thought he was quite clever when he dreamed up this technique, but he was quickly informed that the algorithm is well known and goes by the name of the method of local variations. We will look at the algorithm as applied to a simple control problem, rather than to the calculus of variations, since this is the setting in which it is found in the references. See Cullum [81] and Krylov and Chernousk'ko [164].

Consider the problem of controlling a state variable $x \in R^n$ from an arbitrary point at time S to a fixed point X at a later time T, by a control $u: [S, T] \to R^m$, subject to a differential equation

$$f(t, x(t), x'(t), u(t)) = 0. \tag{5.1}$$

A cost functional

$$J(u) = \varphi(x(S)) + \int_S^T g(t, x(t), u(t)) \, dt$$

is to be minimized over some class $\mathcal{U}$ of admissible controls which are at least bounded and measurable. Here φ and g are continuous real functions which are bounded below.

This problem may be approximated by a decision tree as follows. Choose a positive integer p and divide the interval $[S, T]$ into p equal parts by a partition

$$T = t_0 > t_1 > \cdots > t_p = S.$$

That is,

$$t_k = T - k(T - S)/p.$$

On $[t_1, t_0]$ define $u(t) \equiv u_1$, where each component of u_1 is chosen to lie in $[-p, p]$ and to be an integrable multiple of $1/p$. Let $x(t)$ be the (assumed unique) solution to (5.1) on $[t_1, t_0)$ with $x(t_0) = X$. Define x_1 to be $x(t_1)$.

Proceeding by induction, at the kth step, choose a u_k as above and find $x(t)$ on $[t_k, t_{k-1}]$ from (5.1) and the terminal condition $x(t_{k-1}) = x_{k-1}$. Define $x_k = x(t_k)$. With each step k is associated a cost

$$c_k = \int_{t_k}^{t_{k-1}} g(t, x(t), u_k)\, dt - M/p, \qquad (5.2)$$

where M is chosen to make each c_k nonnegative. For the final step, c_p will have the initial value $\varphi(x_p)$ added to it. This construction may be considered as a decision tree where the total cost to be minimized is

$$C(u_1, \ldots, u_p) = \sum c_k = J(u) - M. \qquad (5.3)$$

Note that, for each p, there are a finite number of paths through the tree, so that $C(u_1, \ldots, u_p)$ can be minimized, yielding a control v_p and a corresponding minimal cost C_p. As p increases, C_p may be expected to decrease. At least for any subsequence of powers of a fixed prime, C_p is nonincreasing, since the controls of each tree are included in the next. To force C_p to be nonincreasing, we add the path corresponding to v_{p-1} to the tree if it is not already included in it. It would be hoped that, as p grows without bound, v_p converges in some norm to an optimal control v, and C_p tends to $I - M$, where

$$I = \inf\{J(u) : u \in \mathcal{U}\}.$$

This is indeed true in many cases. In any case, the sequence of controls $\{v_p\}$ yields smaller and smaller costs as p increases. In most practical problems, this is of value in itself.

The total number of paths through a tree with p intervals as above is $P = (2p^2 + 1)^{mp}$. Thus when m is large, or the controls considered are not uniformly bounded so that large values of p are needed, the tree becomes so unwieldy that evaluation of C_p is impractical even though theoretically possible. For a one-dimensional case ($m = 1$) and $p = 5$, this results in 51^5 paths to be evaluated. Thus efficient strategy for evaluation of C_p is imperative.

One strategy for optimizing decision trees which is well known and has many variants is the branch and bound technique. The author wishes to thank J. Friedenfelds of Bell Laboratories for suggesting background and reference material covering branch and bound algorithms.

The basic method is outlined, with examples, in Hiller and Lieberman [139]. In simplest form, it proceeds as follows. Select a control u^0 of the tree, and the associated cost $C^0 = J(u^0) - M$. A reasonable selection would be $u^0 = v_{p-1}$ or $u^0 \equiv 0$. Then every possible path of the tree is examined. If, after k steps, the accumulated cost $\sum \{c_j : 1 \le j \le k\}$ exceeds or equals C^0, all paths with the same first k branches as the given paths are eliminated from consideration. If the total cost for a path is less than C^0, the control corresponding to this path replaces u^0 and its cost replaces C^0. The algorithm proceeds until all paths have either been eliminated or used to replace u^0 and C^0.

Clearly a good choice of u^0 and efficient strategy for eliminating paths and choosing good paths early in the algorithm can greatly affect the speed and ease of obtaining v_p and C_p. In recent years several such strategies have appeared in the literature. A summary of techniques known in 1966 appears in Lawler and Wood [175].

For simplicity only a one-dimensional variational problem will be considered for the remainder of this section. That is, $m = n = 1$, $S = 0$, $T = 1$, and $f(t, x, x', u) = x' - u$, or

$$x(t) = X + \int_1^t u(\tau)d\tau.$$

For a path $u = (u_1, \ldots, u_p)$ define

$$u_i{}^+ = (u_1, \ldots, u_i + 1/p, u_{i+1}, \ldots, u_p),$$
$$u_i{}^- = (u_1, \ldots, u_i - 1/p, u_{i-1}, \ldots, u_p), \qquad 1 \le i \le p. \qquad (5.4)$$

If u is optimal, both $C(u_i{}^+) - C(u)$ and $C(u_i{}^-) - C(u)$ are nonnegative. If not, letting k be the first i such that $C(u_i{}^+) - C(u)$ or $C(u_i{}^-) - C(u)$ is negative, where $u = u^0$, consider next the path $u_i{}^+$ or $u_i{}^-$, whichever yields the negative difference. This strategy will lead to a strategy u^0 in which all of the above differences are nonnegative and usually in a fairly efficient manner. With convexity imposed on φ and g, this yields v_p. In general, it leads to a sort of discrete "local" optimum and other paths must still be tested.

Example 5.1 Let $m = n = 1$, $g(t, x, u) = (1 + t)u^2$, $\varphi(x) = (x - 1)^2$. Consider integrable controls u with $\|u\|_\infty \le 1$. Then

$$J(u) = (\lambda - 1)^2 + \int_0^1 (1 + t)u^2(t) \, dt,$$

where $\lambda = x(0)$.

The classical Euler equation yields a unique continuous optimal control:

$$u(t) = -(1 + \ln 2)^{-1}(1 + t)^{-1}$$

for which

$$x(t) = (\ln 2 - \ln(1 + t))/(1 + \ln 2),$$

and

$$J(u) = 1/(1 + \ln 2) \approx 0.5906.$$

To obtain some idea of the efficiency of the branch and bound algorithm for this problem, the first few v_p will be given below.

For any p, the cost function here is

$$C(u_1, \ldots, u_p) = (1 + \sum u_k/p)^2 + \sum u_k^2[4p + 1 - 2k]/(2p^2).$$

Letting $n_k = -pu_k$, n_k may take on the values $0, \pm 1, \ldots, \pm p$. Define

$$K(n_1, \ldots, n_p) = 2p^4 C(u_1, \ldots, u_p) = 2(p^2 - \sum n_k)^2 + \sum n_k^2(4p + 1 - 2k).$$

Note that if $n_k < 0$, it may be replaced by 0, reducing K. Also if $i < j$ and $|n_i| > |n_j|$, n_i and n_j may be interchanged, reducing K. Thus it is necessary to consider only n_k such that

$$0 \le n_1 \le \cdots \le n_p \le p.$$

Starting with $p = 1$ and $u^0 = 0$, one obtains immediately $v_1 \equiv 0$, $C_1 = 1$. For $p = 2$, examining the 5 new paths yields $v_2 \equiv -\frac{1}{2}$, $C_2 = 0.625$. For $p = 3$, there are 18 new paths to consider,

$$v_3(t) = \begin{cases} -\frac{2}{3}, & 0 \le t < \frac{1}{3}, \\ -\frac{1}{3}, & \frac{1}{3} \le t < 1, \end{cases}$$

$$C_3 = \frac{49}{81} \approx 0.6049.$$

For $p = 4$, the number of new paths is quite large, so use of u_i^+ and u_i^- from (5.4) becomes helpful. Using $u^0 = V_2$, one obtains in a few steps

$$v_4(t) = \begin{cases} -\frac{1}{2}, & 0 \le t \le \frac{1}{2}, \\ -\frac{1}{4}, & \frac{1}{2} \le t \le 1, \end{cases}$$

$$C_4 = \frac{77}{128} \approx 0.6016.$$

To four-place accuracy, one obtains the following table in a matter of a few minutes of hand calculations.

$p =$	1	2	3	4	5	∞
$C_p =$	1.0000	0.6250	0.6049	0.6016	0.5976	0.5906.

The approximation of a control problem by a finite decision tree in the manner described is always possible. If the optimal paths v_p do not converge to an optimal control, they are at least increasingly good controls, in the sense that $J(v_p)$ does not increase. In many cases they do converge to an optimal control, and $J(v_p)$ converges to the optimum value.

The branch and bound algorithm, supplemented by a consideration of finite differences, may be used to optimize the decision trees. The algorithm is admittedly crude but often works where no other algorithm is known. If numerical integration is needed and a large tree is desired for accuracy, the speed and memory limits of present day computers will be exceeded. However at least a rough approximation by this technique is always possible. At the last step, it may be desirable to smooth v_p or the corresponding x. This may be accomplished by fitting a least squares polynomial or by using an averaging kernel. This often yields a final improvement in C_p.

Should the terminal state and time, X and T, be variable, the decision tree approximation is still possible with T and X as well as $u_1, \ldots, u_p$ to be chosen, thus adding two steps to each tree.

Of course, in applying any of these numerical techniques to the Hamilton–Jacobi equation, an absolute minimum is essential, and the necessary convergence must be established. The reader should check the references for the theorems needed. For further study in numerical control theory, see Polak [216] and Sarachik [235].

Before leaving the subject of numerical solutions of variational problems, we should at least mention a class of techniques based upon "normalization" via contact transformations. These methods attack more directly the Hamiltonian system of ODEs, but may be used to yield extremals and thus solve the variational problem. However, they are generally local in nature, so do not always yield global solutions. For references, see Birkhoff [40], Gustavson [126], and Bernstein [38]. For a FORTRAN program based upon the Birkhoff normalization process, see Chai and Kass [58].

6. NUMERICAL METHODS FOR FIRST-ORDER EQUATIONS

There are a wealth of finite difference schemes which may be applied directly to first-order equations, although the bulk of the emphasis is usually on second-order equations. For an elementary exposition, see Street [249, chapter 10]. For further study in this area, see Ames [8], Fox [112], and Smith [243].

One point that should be noted is, that for the problems that we have been considering, the characteristic slope is bounded. Thus if t and x range over a grid in the one-dimensional case,

$$t = kh, \qquad k = 0, 1, \ldots,$$
$$x = nl, \qquad n = 0, \pm 1, \pm 2, \ldots, \tag{6.1}$$

with h and l fixed positive numbers, the value $u(t, x) = U(m, n)$ depends only upon data in a cone of determinacy around (t, x). It seems then, that the ratio $\rho = l/h$ must be kept within reasonable bounds, if the difference scheme is to approximate a solution of the desired PDE. This is, in fact, the case. With the necessary hypotheses assumed, if ρ is kept properly bounded, and $h \to 0$, the difference scheme will tend to the true solution. But if $h \to 0$ while $\rho \to \infty$, strange results often occur, even to the point of solving a completely different PDE.

The only method we shall consider for global solutions applies to the quasilinear equation

$$u_t + \partial[H(t, x, u)]/\partial x + \psi(t, x, u) = 0 \tag{6.2}$$

subject to

$$u(0, x) = f(x), \qquad x \in \mathbb{R}. \tag{6.3}$$

Suppose that the half plane $t \geq 0$ is covered by a grid as in (6.1), and consider the difference scheme of Lax [179], where (6.2) is replaced by the following system of finite difference equations:

$$\frac{u_n^{k+1} - (u_{n-1}^k + u_{u+1}^k)/2}{h} + \frac{H(kh, (n+1)l, u_{n+1}^k) - H(kh, (n-1)l, u_{n-1}^k)}{2}$$

$$+ \psi(kh, (n+1)l, u_{n+1}^k) = 0, \tag{6.4}$$

where $u_i{}^j$ denotes $u(jh, il)$. In solving these equations, the value of u_n^{k+1} is computed from the values with $t = kh$, so that the evolution in time is easy to calculate. That this scheme does indeed lead to a solution of (6.2) and (6.3) was shown by Oleinik [203]. We shall outline her development briefly.

First assume that H_x, H_u, H_{xx}, H_{xu}, H_{uu}, ψ_x, and ψ_u are continuous and bounded for bounded u, all (t, x) with $t \geq 0$. Assume also that $H_{uu} \geq \mu > 0$ for $0 \leq t \leq \tau$, and bounded u, where τ and μ are positive constants, and $H_{uu} > 0$ for $t > 0$. In addition to (6.4), require that

$$u_n{}^0 = f(nl), \tag{6.5}$$

and $\|f\|_\infty \leq m$. Assume that there is a constant M and a continuously differentiable function V such that

$$\max_{\substack{|u| \leq v \\ t \geq 0}} |H_x + \psi| < V(v), \, V'(v) \geq 0,$$

and

$$\int_m^M \frac{dv}{V(v) + \alpha} \geq T,$$

for some $\alpha > 0$. Denote the region where $|u| \leq M$ and $t \geq 0$ by Ω, and let $A = \max_\Omega |H_u|$.

Now consider, for $t \geq 0$, the family of functions constructed from $u_n{}^k$ as follows. For (t, x) satisfying

$$kh \leq t < (k+1)h, \qquad nl \leq x < (n+2)l,$$

and for $k - n$ even, define $U_{hl} = u_n{}^k$. Thus $U_{hl}(t, x)$ is defined for all $t \geq 0$ and all $x \in \mathbb{R}$ and agrees with $u_n{}^k$ on the grid points (6.1). Oleinik shows the convergence of Lax's difference scheme in the following important theorem. The proof, which we omit, is a long but beautiful piece of analysis.

THEOREM 6.1   *Oleinik's Theorem*   If l^2/h remains bounded, $Ah/l < 1$, then it is possible to select from the family of functions $\{U_{hl}\}$ an infinite sequence $\{U_{hl}^i\}$ such that for h and $l \to 0$, $i \to \infty$, and any $X > 0$,

$$\int_{-X}^{X} |U_{hl}^i(t, x) - u(t, x)|\, dx \to 0.$$

The limit function $u(t, x)$ is measurable, $|u(t, x)| \leq M$ for $t \geq 0$, and

$$\int_{0}^{T} \int_{-X}^{X} |U_{hl}^i(t, x) - u(t, x)|\, dx\, dt \to 0 \qquad \text{as} \quad i \to \infty.$$

The function u is a solution of (6.2) and (6.3) in the sense that

$$\iint_{D} [g_t u(t, x) + g_x H(t, x, u) - g\psi(t, x, u)]\, dx\, dt$$

$$+ \int_{-\infty}^{\infty} g(0, x) f(x)\, dx = 0$$

for any thrice continuously differentiable function g in $t \geq 0$ with compact support which is zero at $t = T$, if $l^2/h \to 0$ and if

$$\int_{-\infty}^{\infty} g(0, x)[f(x) - U_{hl}^i(0, x)]\, dx \to 0 \qquad \text{as} \quad i \to \infty. \qquad \blacksquare$$

Besides the difference schemes, Douglis [87] proved existence of a solution by a mixed difference-differential scheme. In his method, which we shall not detail here, the x-variable in $\mathbb{R}^n$ is broken up into a grid as in a difference scheme but the t-values are determined, for each x-grid point, by a differential equation. Thus any numerical method for an ODE initial value problem, such as Runge–Kutta or predictor-corrector techniques, can be used in conjunction with the x-difference scheme to yield a numerical solution for very general Cauchy problems. Douglis' convergence theorem and proof are another beautiful piece of analysis and are heartily recommended to the interested reader. For an excellent introduction to the numerical solution of ODEs, see Shampine *et al.* [239].

7. ARTIFICIAL VISCOSITY

Fleming's work based upon the method of "artificial viscosity," or higher-order parabolic equations, has already been mentioned. In this section, we will outline it in a little more detail. Fleming actually worked with $s = -t$, and $F = -H$, solving the so-called backward Cauchy problem. The reason for this is that the stochastic arguments used in solving the second-order parabolic equations more naturally lend themselves to the backward problem, although in substance there is little difference.

Thus Fleming considers the problem

$$u_s + F(s, x, u_x) = 0, \qquad T_0 \leq s \leq T \tag{7.1}$$

subject to

$$u(T, x) = f(x). \tag{7.2}$$

He defines a solution to be a bounded, Lipschitzian function u satisfying (7.1) almost everywhere in $D = [T_0, T] \times \mathbb{R}^n$. As we have seen such a solution exists with (7.2) holding identically; that is, no fussing about the boundary approaches is necessary here. He obtains the solution as a limit as $\epsilon \to 0^+$ of solutions of

$$u_s^\epsilon + (\epsilon^2/2)\Delta_x u^\epsilon + F(s, x, u_x^\epsilon) = 0, \qquad T_0 \leq s \leq T, \tag{7.3}$$

with

$$u^\epsilon(T, x) = f^\epsilon(x), \tag{7.4}$$

where $f^\epsilon \to f$ as $\epsilon \to 0^+$ and Δ_x denotes the Laplace operator in the space variables (x).

Fleming assumes that $F(s, x, p)$ is of class $C^{(3)}$ and

(1) $|F_s| + |F_p| + |F_{x_i p_j}| + |F_{s p_j}| \leq Q(|p|),$

$$\gamma(|p|)|\lambda|^2 \leq -\lambda^T F_{pp} \lambda \leq Q(|p|)|\lambda|^2,$$

for all $\lambda \in \mathbb{R}^n$, where $\gamma(v)$ and $Q(v)$ are positive functions, respectively nonincreasing and nondecreasing in v.

(2) $\lim\limits_{|p| \to \infty} F(s, x, p)/|p| = -\infty.$

Given $r > 0$ there exists k_r such that $|F_p| \leq r$ implies

$$|p| \leq k_r.$$

(3) $F(s, x, p) \leq C$ for some C.
(4) $|F_x| \leq c_1(F - pF_p) + c_2$ for some positive c_1, c_2.

These assumptions are essentially those we have seen that were needed before with slightly different notation. Fleming also assumes that f is of class C^1, is bounded, and Lipschitzian. Defining

$$L(s, x, y) = \max_p [F(s, x, p) - yp],$$

Fleming considers the variational problem of minimizing

$$J(\alpha) = \int_s^T L(\tau, \alpha, \alpha')\, d\tau + f(\alpha(T))$$

among class C^1 curves with $\alpha(s) = x$, and establishes that

$$u(s, x) = \min_\alpha J(\alpha)$$

is a solution of (7.1) and (7.2). He also shows that at almost every point (s, x) there is a unique minimizing curve and calls such points regular points. He points out a theorem of Kuznetzov and Siskin [166], stating that u is differentiable and satisfies (7.1) at precisely the regular points.

Now let f^ϵ be a class C^∞ function (for each $\epsilon > 0$), with $|f^\epsilon| \le M_0$, $|f_x^\epsilon| \le N_0$, where M_0 and N_0 are the corresponding bounds for f, such that the higher order derivatives are bounded by constants depending upon ϵ, and as $\epsilon \to 0^+$, f^ϵ and f_x^ϵ tend uniformly to f and f_x. (Use mollified functions.)

The problems (7.3) and (7.4) can be solved by considering a stochastic variational problem. Toward this end, consider n-dimensional stochastic processes ξ on $[s, T]$ of the form

$$\xi(t) = \eta(t) + \epsilon[w(t) - w(s)], \tag{7.5}$$

where η is a process with sample paths of class C^1, $\eta(s) = x$, and w is an n-dimensional Brownian motion. Require also that ξ be nonanticipative; that is, the random variables $\xi(t)$ for $r \le t$ are independent of the Brownian increments for times $\ge t$. Fleming then proceeds to minimize

$$J^\epsilon(\xi) = E\left\{ \int_s^T L(t, \xi, \eta')\, dt + f^\epsilon(\xi(T)) \right\}, \tag{7.6}$$

where $E\{\cdot\}$ denotes the expected value. Fleming actually restricts his stochastic processes ξ even further, then shows that the minimum is indeed attained. Denoting this minimum by $u^\epsilon(s, x)$, which solves (7.3) and (7.4), Fleming obtains the following theorem.

THEOREM 7.1 (*Fleming*) As $\epsilon \to 0^+$:

(a) $u^\epsilon(s, x)$ tends uniformly to $u(s, x)$.
(b) $u_x^\epsilon(s, x)$ tends to $u_x(s, x)$ at every regular point (s, x).   ∎

Fleming also shows that u above satisfies the requirements of the uniqueness theorem of Kruzhkov or that of Douglis.

Thus for numerical work, the stochastic theory may be abandoned, but the crucial point is that the "correct" u may be obtained as a limit as $\epsilon \to 0^+$ of solutions of (7.3) and (7.4). Thus any methods which apply to these second-order parabolic equations may be applied to the Hamilton–Jacobi equation.

Again, the pertinent techniques are primarily difference schemes, which we shall not detail. But these may be found easily in the literature. See, for example, Street [249, chapter 10] and Ames [8, chapter 2].

Bibliography

[1] Abraham, R., "Foundations of Classical Mechanics." Benjamin, New York, 1967.

[2] Aizawa, S., A semigroup treatment of the Hamilton–Jacobi equation in one space variable, *Hiroshima Math. J.* **3** (1973), 367–386.

[3] Aizawa, S., and Kikuchi, N., A mixed initial and boundary-value problem for the Hamilton–Jacobi equation in several space variables, *Funkcial. Ekvac.* **9** (1966), 139–150.

[4] Aizenman, M., Gallavotti, G., Goldstein, S., and Lebowitz, J. L., Stability and equilibrium states of infinite classical systems, *Comm. Math. Phys.* **48** (1976), 1–14.

[5] Akhiezer, N. I., "The Calculus of Variations" (A. H. Frink, translator, G. Springer ed.), Ginn (Blaisdell), Boston Massachusetts, 1962.

[6] Al'ber, Y. I., The problem of the minimization of smooth functionals by gradient methods, *USSR Compt. Math. Math. Phys.* **11** (1971), 263–272.

[7] Ames, W. F., "Nonlinear Partial Differential Equations in Engineering." Academic Press, New York, 1965.

[8] Ames, W. F., "Numerical Methods for Partial Differential Equations." Barnes and Noble, New York, 1967.

[9] Ames, W. F., "Nonlinear Partial Differential Equations in Engineering," Vol. II. Academic Press, New York, 1972.

[10] Ames, W. F., Implicit Ad Hoc methods for nonlinear partial differential equations, *J. Math. Anal. Appl.* **42** (1973), 20–28.

[11] Anastassov, A. H., Relativity and quantization, *Il Nuovo Cimento* **14** (1973), 52–56.

[12] Angel, E., and Bellman, R., "Dynamic Programming and Partial Differential Equations." Academic Press, New York, 1972.

[13] Anselone, P. M. (ed.), "Nonlinear Integral Equations." Univ. of Wisconsin Press, Madison, Wisconsin, 1964.

[14] Antosiewicz, H. A., "International Conference on Differential Equations." Academic Press, New York, 1975.

[15] Aseltine, J. A., "Transform Methods in Linear System Analysis." McGraw-Hill, New York, 1958.

[16] Assad, N. A., and Kirk, W. A., Fixed point theorems for set-valued mappings of contractive type, *Pacific J. Math.* **43** (1972), 553–562.

[17] Auslander, L., and MacKenzie, R. E., "Introduction to Differentiable Manifolds." McGraw-Hill, New York, 1963.

[18] Bailey, P. B., Shampine, L. F., and Waltman, P. E., "Nonlinear Two Point Boundary Value Problems." Academic Press, New York, 1968.

[19] Ballou, D. P., Solutions to nonlinear hyperbolic Cauchy Problems without convexity conditions, *Trans. Amer. Math. Soc.* **152** (1970), 441–460.

[20] Barros-Neto, J., "An Introduction to the Theory of Distributions." Dekker, New York, 1973.

[21] Bautin, N. N., Qualitative investigation of a dynamic system, *Appl. Math. Mech.* **36** (1972), 388–394.

[22] Belinfante, J. G. F., and Kolman, B., Lie Groups and Lie Algebras with Applications and Computational Methods. S.I.A.M., Philadelphia, Pennsylvania, 1972.

[23] Bell, D. J., and Jacobson, D. H., "Singular Optimal Control Problems." Academic Press, New York, 1975.

[24] Bellman, R., "Dynamic Programming." Princeton Univ. Press, Princeton, New Jersey, 1957.

[25] Bellman, R., "Adaptive Control Processes: A Guided Tour." Princeton Univ. Press, Princeton, New Jersey, 1961.

[26] Bellman, R., "Introduction to the Mathematical Theory of Control Processes," Vol. I. Academic Press, New York, 1967.

[27] Bellman, R., "Modern Elementary Differential Equations." Addison-Wesley, Reading, Massachusetts, 1968.

[28] Bellman, R., "Methods of Nonlinear Analysis," Vol. I. Academic Press, New York, 1970.

[29] Benton, S. H., A general space-time boundary value problem for the Hamilton–Jacobi equation, *J. Differential Equations* **11** (1972), 425–435.

[30] Benton, S. H., Global Solutions of Hamilton–Jacobi Boundary Value Problems by Variational Methods. Tulane Univ. Dissertation, New Orleans, Louisiana, 1972.

[31] Benton, S. H., Global variational solutions of Hamilton–Jacobi boundary value problems, *J. Differential Equations* **13** (1973), 468–480.

[32] Benton, S. H., Necessity of the Compatibility Condition for Global Solutions of Hamilton–Jacobi Boundary Value Problems (paper to be submitted for publication).

[33] Berg, P. W., and McGregor, J. L., "Elementary Partial Differential Equations." Holden-Day, San Francisco, California, 1966.

[34] Berkovitz, L. D., Variational methods in problems of control and programming, *J. Math. Anal. Appl.* **3** (1961), 145–169.

[35] Berkovitz, L. D., On Control Problems with Bounded State Variables. Rand Corp. Memo no. RM 3207–PR, 1962, ASTIA AD 278262.

[36] Berkovitz, L. D., "Optimal Control Theory." Springer-Verlag, Berlin and New York, 1974.

[37] Berkovitz, L. D., and Dreyfus, S. E., A dynamic programming approach to the non-parametric problem in the calculus of variations, *J. Math. Mech.* **15** (1966), 83–100.

[38] Bernstein, I. B., Finite Time Stability of Periodic Solutions of Hamiltonian Systems. NASA TM X-53478, Huntsville, Alabama, 1966.

[39] Bers, L., John, F., and Schecter, M., "Partial Differential Equations." Wiley (Interscience), New York, 1964.

[40] Birkhoff, G. D., "Dynamical Systems." Amer. Math. Soc., Providence, Rhode Island, 1927.

[41] Birkhoff, G. D., "Collected Mathematical Papers, George David Birkhoff," Vol. II, Dynamics (continued)/Physical Theories. Dover, New York, 1968.

[42] Bliss, G. A., "Lectures on the Calculus of Variations." Univ. of Chicago Press, Chicago, Illinois, 1946.

[43] Bluman, G. W., and Cole, J. D., "Similarity Methods for Differential Equations." Springer-Verlag, Berlin and New York, 1974.

[44] Boltyanskii, V. G., "Mathematical Methods of Optimal Control" (K. N. Trirogoff, translator). Holt, New York, 1971.

[45] Bolza, O., "Calculus of Variations," 2nd ed. Chelsea, New York, 1904.

[46] Bourbaki, N., "Éléments de Mathématiques." Hermann, Paris, 1939–48.

[47] Brandt, U., and Leschke, H., Weak self-consistent approximation scheme in many-body systems, Z. Phys. **260** (1973), 147–156.

[48] Braun, M., "Differential Equations and Their Applications." Springer-Verlag, Berlin and New York, 1975.

[49] Brent, R., Winograd, S., and Wolfe, P., Optimal iterative processes for root-finding, Numer. Math. **20** (1973), 327–341.

[50] Burch, B. C., A Semigroup Approach to the Hamilton–Jacobi Equation, Tulane Univ. dissertation, New Orleans, Louisana, 1975.

[51] Burch, B. C., A semigroup treatment of the Hamilton–Jacobi equation in several space variables, J. Differential Equations (to appear).

[52] Caratheodory, C., "Calculus of Variations and Partial Differential Equations of the First Order," Part I. Holden-Day, San Francisco, California, 1965.

[53] Carroll, R. W., "Abstract Methods in Partial Differential Equations." Harper, New York, 1969.

[54] Casti, J., and Kalaba, R., "Imbedding Methods in Applied Mathematics." Addison-Wesley, Reading, Massachusetts, 1973.

[55] Cesari, J., Existence theorems for weak and usual optimal solutions in Lagrange problems with unilateral constraints, I and II, Trans. Amer. Math. Soc. **124** (1966), 369–430.

[56] Cesari, L., Existence theorems for problems of optimization with distributed and boundary controls, Actes Congres Intern. Math. **3** (1971), 157–161.

[57] Cesari, L., Hale, J. K., and LaSalle, J. P., "Dynamical Systems, An International Symposium I." Academic Press, New York, 1976.

[58] Chai, W. A., and Kass, S., Birkhoff Normalization Process Program for Time-Independent Hamiltonian Systems. AFOSR Scientific Rep. No. AFOSR 67–0123, General Precision Aerospace, Little Falls, New Jersey, 1966.

[59] Chern, S. S., Differentiable Manifolds. Special Notes, Mathematics 243, Univ. of Chicago, Chicago, Illinois, 1959.

[60] Chester, C. R., "Techniques in Partial Differential Equations." McGraw-Hill, New York, 1971.

[61] Choquet-Bruhat, Y., "Problems and Solutions in Mathematical Physics" (C. Peltzer, translator). Holden-Day, San Francisco, California, 1967.

[62] Churchill, R. C., and Rod, D. L., Pathology in dynamical systems, I: General theory, J. Differential Equations **21** (1976), 39–65.

[63] Churchill, R. C., and Rod, D. L., Pathology in dynamical systems, II: Applications, J. Differential Equations **21** (1976), 66–112.

[64] Churchill, R. V., "Complex Variables and Applications," 2nd ed., McGraw-Hill, New York, 1960.

[65] Cioranescu, I., Sur les semigroupes ultra-distributions, J. Math. Anal. Appl. **41** (1973), 539–544.

[66] Cohen, A. M., "Numerical Analysis." Halsted Press, New York, 1973.

[67] Cole, J. D., On a quasilinear parabolic equation occuring in aerodynamics, Quart. Appl. Math. **9** (1951), 225–236.

[68] Cole, J. D., "Pertubation Methods in Applied Mathematics." Ginn (Blaisdell), Boston, Massachusetts, 1968.

[69] Conway, E. D., Generalized solutions of linear differential equations with discontinuous coefficients and the uniqueness question for multidimensional quasilinear conservation laws, *J. Math. Anal. Appl.* **18** (1967), 238–251.

[70] Conway, E. D., Stochastic equations with discontinuous drift, *Trans. Amer. Math. Soc.* **157** (1971), 235–245.

[71] Conway, E. D., On the total variation of solutions of parabolic equations, *Indiana Univ. Math. J.* **21** (1971), 493–503.

[72] Conway, E. D., The Formation and Decay of Shocks for a Conservation Law in Several Dimensions. Tulane Univ. Preprints, New Orleans, Louisiana, 1975.

[73] Conway, E. D., and Hopf, E., Hamilton's theory and generalized solutions of the Hamilton–Jacobi equation, *J. Math. Mech.* **13** (1964), 939–986.

[74] Conway, E. D., and Smoller, J., Global solutions of the Cauchy problem for quasilinear first order equations in several space variables, *Comm. Pure Appl. Math.* **19** (1966), 95–105.

[75] Courant, R., and Friedrichs, K. O., "Supersonic Flow and Shock Waves." Wiley (Interscience), New York, 1948.

[76] Courant, R., and Hilbert, D., "Methods of Mathematical Physics," Vol. I. Wiley (Interscience), New York, 1953.

[77] Courant, R., and Hilbert, D., "Methods of Mathematical Physics," Vol. II, Partial Differential Equations. Wiley (Interscience), New York, 1962.

[78] Crandall, M. G., The semigroup approach to first order quasilinear equations in several space variables, *Israel J. Math.* **12** (1972), 108–132.

[79] Crandall, M. G., and Liggett, T. M., A theorem and a counter-example in the theory of semigroups of nonlinear transformations, *Trans. Amer. Math. Soc.* **160** (1971), 263–278.

[80] Crandall, M. G., and Liggett, T. M., Generation of semi-groups of nonlinear transformations on general Banach spaces, *Amer. J. Math.* **93** (1971), 265–298.

[81] Cullum, J., Finite dimensional approximations of state-constrained continuous optimal control problems, *SIAM J. Control* **10** (1972), 649–670.

[82] Denman, H. H., and Buch, L. H., Solution of the Hamilton–Jacobi equation for certain dissipative classical mechanical systems, *J. Math. Phys.* **14** (1973), 326–329.

[83] Devaney, R. L., Homoclinic orbits in Hamiltonian systems, *J. Differential Equations* **21** (1976), 431–438.

[84] Dieudonné, J., "Foundations of Modern Analysis." Academic Press, New York, 1960.

[85] Diperna, R. J., Global solutions to a class of nonlinear hyperbolic systems of equations, *Comm. Pure Appl. Math.* **16** (1973), 1–28.

[86] Douglis, A., An ordering principle and generalized solutions of certain quasi-linear partial differential equations, *Comm. Pure Appl. Math.* **12** (1959), 87–112.

[87] Douglis, A., Solutions in the large for multi-dimensional non-linear partial differential equations of first order, *Ann. Inst. Fourier Grenoble* **15** (1965), 1–35.

[88] Douglis, A., On weak solutions of non-linear partial differential equations with real characteristics, *Math. Ann.* **163** (1966), 351–358.

[89] Douglis, A., Layering methods for nonlinear partial differential equations of first order, *Ann. Inst. Fourier Grenoble* **22** (1972), 141–227.

[90] Dunford, N., and Schwartz, J. T., "Linear Operators", Part I. Wiley (Interscience), New York, 1958.

[91] Dunkl, C. F., and Ramirez, D. E., "Topics in Harmonic Analysis." Appleton, New York, 1971.

[92] Eisenhart, L. P., "Continuous Groups of Transformation." Dover, New York, 1961.

[93] Elliott, R. J., Quasilinear resolutions of non-linear equations, *Manuscripta Math.* **12** (1974), 399–410.

[94] Elliott, R. J., Boundary Value Problems for Non-linear Partial Differential Equations, *in* "Global Analysis and Its Applications," Vol. II, pp. 145–149. Int. At. Energy Agency, Vienna, 1974.

[95] Elliott, R. J., and Friedman, A., A note on generalized pursuit-evasion games, *SIAM J. Control* **13** (1975), 105–109.

[96] Elliott, R. J., and Kalton, N. J., The Existence of Value in Differential Games. Amer. Math. Soc. Memoir 126, Providence, Rhode Island, 1972.

[97] Elliott, R. J., and Kalton, N. J., The existence of value in differential games of pursuit and evasion, *J. Differential Equations* **12** (1972), 504–523.

[98] Elliott, R. J., and Kalton, N. J., Cauchy problems for certain Isaacs-Bellman equations and games of survival, *Trans. Amer. Math. Soc.* **198** (1974), 45–72.

[99] Elliott, R. J., and Kalton, N. J., Boundary value problems for nonlinear partial differential operators, *J. Math. Anal. Appl.* **46** (1974), 228–241.

[100] El'sgol'ts, L. E., "Introduction to the Theory of Differential Equations with Deviating Arguments." Holden-Day, San Francisco, California, 1966.

[101] Feltus, E. E., Mixed Problems for the Hamilton–Jacobi Equation. Tulane Univ. Dissertation, New Orleans, Louisana, 1975.

[102] Fenchel, W., On conjugate convex functions, *Can. J. Math.* **1** (1949), 73–77.

[103] Filippov, A. F., Differential equations with discontinuous right-hand side, *Amer. Math. Soc. Transl. Ser. 2,* **42** (1964), 199–231.

[104] Finkelstein, R. J., "Nonrelativistic Mechanics." Addison-Wesley, Reading, Massachusetts, 1973.

[105] Fitzpatrick, P. M., "Principles of Celestial Mechanics." Academic Press, New York, 1970.

[106] Fleming, W. H., The Cauchy problem for degenerate parabolic equations, *J. Math. Mech.* **13** (1964), 987–1008.

[107] Fleming, W. H., "Functions of Several Variables." Addison-Wesley, Reading, Massachusetts, 1965.

[108] Fleming, W. H., Duality and a priori estimates in Markovian optimization problems, *J. Math. Anal. Appl.* **16** (1966), 254–279.

[109] Fleming, W. H., The Cauchy Problem for a nonlinear first order partial differential equation, *J. Differential Equations* **5** (1969), 515–530.

[110] Fleming, W. H., Stochastic Control for Small Noise Intensities, Tech. Rep. 70–1. Center for Dynamical Systems, Brown Univ., Providence, Rhode Island, 1970.

[111] Forsyth, A. R., "Theory of Differential Equations," Vols 5 and 6. Dover, New York, reprinted 1959.

[112] Fox, L., "Numerical Solution of Ordinary and Partial Differential Equations." Addison-Wesley, Reading, Massachusetts, 1962.

[113] Friedman, A., "Partial Differential Equations." Holt, New York, 1969.

[114] Garabedian, P. R., "Partial Differential Equations." Wiley, New York, 1964.

[115] Gelfand, I. M., and Fomin, S. V., "Calculus of Variations" (R. A. Silverman, translator). Prentice-Hall, Englewood Cliffs, New Jersey, 1963.

[116] Giacaglia, G. E. O., "Perturbation Methods in Non-Linear Systems." Springer-Verlag, Berlin and New York, 1972.

[117] Gilbert, R. P., "Function Theoretic Methods in Partial Differential Equations." Academic Press, New York, 1969.

[118] Glimm, J., and Lax, P. D., "Decay of Solutions of Systems of Hyperbolic Conservation Laws." Courant Inst., New York, 1969.

[119] Goldstein, J. A., Second order Ito processes, *Nagoya Math. J.* **36** (1969), 27–63.
[120] Goldstein, J. A., Semigroups of Operators and Abstract Cauchy Problems. Tulane Univ. Lecture Notes, New Orleans, Louisana, 1970.
[121] Goldstein, J. A., Nonlinear Semigroups and Nonlinear Partial Differential Equations. Dept. of Math., Tulane Univ., New Orleans, Louisana 1975.
[122] Goldstein, H., "Classical Mechanics." Addison-Wesley, Reading, Massachusetts, 1959.
[123] Gossick, B. R., "Hamilton's Principle and Physical Systems." Academic Press, New York, 1967.
[124] Greenstadt, J. L., A ricocheting gradient method for nonlinear optimization, *SIAM J. Appl. Math.* **14** (1966), 429–445.
[125] Guggenheimer, H. W., "Differential Geometry." McGraw-Hill, New York, 1963.
[126] Gustavson, F., On Constructing Formal Integrals of a Hamiltonian System near an Equilibrium Point. IBM Res. Rep. RC-1556, Yorktown Heights, New York, 1965.
[127] Hajek, O., "Pursuit Games, An Introduction to the Theory and Application of Differential Games of Pursuit and Evasion." Academic Press, New York, 1975.
[128] Hall, D. W., and Spencer, G. L., "Elementary Topology." Wiley, New York, 1955.
[129] Hall, M. Jr., "The Theory of Groups." Macmillan, New York, 1959.
[130] Halmos, P. R., "Measure Theory." Van Nostrand–Reinhold, Princeton, New Jersey, 1950.
[131] Harten, A., Hyman, J. M., and Lax, P. D., On finite-difference approximations and entropy conditions for shocks, *Comm. Pure Appl. Math.* **29** (1976), 297–322.
[132] Hartman, P., "Ordinary Differential Equations." Wiley, New York, 1964.
[133] Heffes, H., and Sarachik, P. E., Uniform approximation of linear systems, *Bell Syst. Tech. J.* **48** (1969), 209–231.
[134] Henrici, P., "Discrete Variable Methods in Ordinary Differential Equations." Wiley, New York, 1962.
[135] Henrici, P., "Elements of Numerical Analysis." Wiley, New York, 1964.
[136] Hermann, R., "Lie Algebras and Quantum Mechanics." Benjamin, New York, 1970.
[137] Hestenes, M. R., "Calculus of Variations and Optimal Control Theory." Wiley, New York, 1966.
[138] Hilbert, D., "Grundzuge einer Allgemeinen Theorie der Linearen Integralgleichungen." Chelsea, New York, 1953.
[139] Hiller, F. S., and Lieberman, G. J., "Introduction to Operations Research." Holden-Day, San Francisco, California, 1967.
[140] Hochschild, G., "The Structure of Lie Groups." Holden-Day, San Francisco, California, 1965.
[141] Hochstadt, H., "The Functions of Mathematical Physics." Wiley (Interscience), New York, 1971.
[142] Hocking, J. G., and Young, G. S., "Topology." Addison-Wesley, Reading, Massachusetts, 1961.
[143] Hopf, E., The partial differential equation $u_t + u\,u_x = \mu u_{xx}$, *Comm. Pure Appl. Math.* **3** (1950), 201–230.
[144] Hopf, E., Generalized solutions of non-linear equations of first order, *J. Math. Mech.* **14** (1965), 951–974.
[145] Horvath, J., An introduction to distributions, *Amer. Math. Monthly* **77** (1970), 227–240.
[146] Hu, S. T., "Elements of General Topology." Holden-Day, San Francisco, California, 1964.
[147] Hubbard, B., "Numerical Solution of Partial Differential Equations," Vol. III. Academic Press, New York, 1976.
[148] Hurewicz, W., "Lectures on Ordinary Differential Equations." M.I.T. Press, Cambridge, Massachusetts, 1958.

[149] Hussain, M., Hamilton–Jacobi Theorem in Group Variables, *J. Appl. Math. Phys.* (*ZAMP*) **27** (1976), 285–287.

[150] Intriligator, M. D., "Mathematical Optimization and Economic Theory." Prentice-Hall, Englewood Cliffs, New Jersey, 1971.

[151] John, F., "Partial Differential Equations." Courant Inst., New York, 1953.

[152] John, F., "Partial Differential Equations," 2nd ed. Springer-Verlag, Berlin and New York, 1975.

[153] Kato, T., Nonlinear semigroups and evolution equations, *J. Math. Soc. Japan* **19** (1967), 508–520.

[154] Keller, H. B., "Numerical Methods for Two-Point Boundary-Value Problems." Ginn (Blaisdell), Boston, Massachusetts, 1968.

[155] Kelley, J. L., and Namioka, I., "Linear Topological Spaces." Van Nostrand–Reinhold, Princeton, New Jersey, 1963.

[156] Kilmister, C. W., and Pirani, F. A. E., Ignorable coordinates and steady motion in classical mechanics, *Proc. Cambridge Philos. Soc.* **61** (1965), 211.

[157] Knill, R. J., Fixed points of uniform contractions, *J. Math. Anal. Appl.* **12** (1965), 449–455.

[158] Knill, R. J., Cones, products and fixed points, *Fund. Math.* **60** (1967), 35–46.

[159] Kolman, B. (ed.), Lie algebras, applications and computational methods, *SIAM J. Appl. Math.* **25** (1973), 163–323.

[160] Kolupanova, G. A., On a differential equation of hyperbolic type in Banach space, *Sov. Math. Dokl.* **13** (1972), 1663–1667.

[161] Kopal, Z., "Figures of Equilibrium of Celestial Bodies." Univ. of Wisconsin Press, Madison, Wisconsin, 1960.

[162] Kothe, G., "Topologische Lineare Raume." Vol. I. Springer, New York, 1960.

[163] Kruzkov, S. N., Generalized solutions of nonlinear first order equations with several independent variables, *Mat. Sb.* **72**, English transl. in *Math. USSR-Sb.* **1** (1967), 93–116.

[164] Krylov, I. A., and Chernous'Ko, F. L., Solution of problems of optimal control by the method of local variations, *U.S.S.R. Comp. Math. and Math. Phys.* **6** (1966), 12–31.

[165] Kuratowski, C., "Introduction to Set Theory and Topology." Pergamon, Oxford, 1962.

[166] Kuznetsov, N. N., and Siskin, A. A., On a many dimensional problem in the theory of quasilinear equations, *Z. Vycisl. Mat. Mat. Fiz.* **4** (1964), 192–205.

[167] Kuznetsov, N. N., and Rozhdestvenskii, B. L., The solution of Cauchy's problem for a system of quasi-linear equations in many independent variables, *Zh. Vych. Mat.* **1** (1961), 217–223.

[168] Lanczos, C., "The Variational Principles of Mechanics." Univ. Toronto Press, Toronto, 1949.

[169] Lanczos, C., "Applied Analysis." Prentice-Hall, Englewood Cliffs, New Jersey, 1956.

[170] Lang, S., "Real Analysis." Addison-Wesley, Reading, Massachusetts, 1969.

[171] Lang, S., "Differential Manifolds." Addison-Wesley, Reading, Massachusetts, 1972.

[172] Langer, R. E. (ed.), "Boundary Value Problems in Differential Equations." Univ. Wisconsin Press, Madison, Wisconsin, 1960.

[173] Langer, R. E. (ed.), "Electromagnetic Waves." Univ. of Wisconsin Press, Madison, Wisconsin, 1962.

[174] Lavi, A., and Vogl, T. P., "Recent Advances in Optimization Techniques." Wiley, New York, 1966.

[175] Lawler, E. L., and Wood, D. E., Branch-and-bound methods: A survey, *Operations Res.* **14** (1966), 699–719.

[176] Lax, P. D., "Partial Differential Equations." Courant Inst., New York, 1951.

[177] Lax, P. D., Nonlinear hyperbolic equations, *Comm. Pure Appl. Math.* **6** (1953), 231–258.

[178]   Lax, P. D., The initial value problem for nonlinear hyperbolic equations in two inde-
        pendent variables, *Ann. Math. Stud.* **33** (1954), 211–229.

[179]   Lax, P. D., Weak solutions of nonlinear hyperbolic equations and their numerical
        computation, *Comm. Pure Appl. Math.* **7** (1954), 159–193.

[180]   Lax, P. D., Hyperbolic systems of conservation laws II, *Comm. Pure Appl. Math.* **10**
        (1957), 537–566.

[181]   Lax, P. D., "Theory of Function of a Real Variable." Courant Inst., New York, 1959.

[182]   Lax, P. D., Nonlinear hyperbolic systems of conservation laws, *In* "Nonlinear Prob-
        lems." Univ. Wisconsin Press, Madison, Wisconsin, 1963.

[183]   Lax, P. D., Numerical solution of partial differential equations, *Amer. Math. Monthly*
        72 (1965), 74–84.

[184]   Lax, P. D., and Nirenberg, L., On stability for difference schemes; a sharp form of
        Garding's inequality, *Comm. Pure Appl. Math.* **19** (1966), 473–492.

[185]   Lee, E. B., and Markus, L., "Foundations of Optimal Control Theory." Wiley, New
        York, 1967.

[186]   Lee, E. S., "Quasilinearization and Invariant Imbedding, with Applications to Chemical
        Engineering and Adaptive Control." Academic Press, New York, 1968.

[187]   MacLane, S., Hamiltonian mechanics and geometry, *Amer. Math. Monthly* **77** (1970),
        570–586.

[188]   Mangasarian, O. L., Robinson, S. M., and Meyer, R. R., "Nonlinear Programming 2."
        Academic Press, New York, 1975.

[189]   Mansfield, M. J., "Introduction to Topology." Van Nostrand–Reinhold, Princeton,
        New Jersey, 1963.

[190]   Maslennikova, V. N., "Explicit Representation of the Solution of the Cauchy Problem
        and Estimates in L_p for Hyperbolic System" (transl. from *Sibirskii Mat. Z.* **13** (1972),
        612–629). Plenum Press, New York, 1972.

[191]   McKelvey, R., "Lectures on Ordinary Differential Equations." Academic Press, New
        York, 1970.

[192]   McShane, E. J., Stochastic integrals and stochastic functional equations, *SIAM J. Appl.
        Math.* **17** (1969), 287–306.

[193]   McShane, E. J., A Riemann-type Integral that Includes Lebesgue–Stieltjas, Bochner
        and Stochastic Integrals. Memoirs AMS, No. 88, AMS, Providence, Rhode Island,
        1969.

[194]   Mendelson, B., "Introduction to Topology." Allyn and Bacon, Boston, Massachusetts,
        1962.

[195]   Mikusinski, J., On partial derivatives, *Bull. Acad. Pol. Sci. Ser. Sci. Math. Astr. Phys.*
        20 (1972), 941–944.

[196]   Mizohata, S., "The Theory of Partial Differential Equations." Cambridge Univ. Press,
        London and New York, 1973.

[197]   Moyer, H. G., Discontinuous Variational Problems. Grumman Res. Dept. Memo. no.
        RM-252J, Grumman Aircraft Eng. Corp., Bethpage, New York, 1964.

[198]   Natanson, I. P., "Theory of Functions of a Real Variable" (L. F. Boron, translator,
        E. Hewitt, ed.), Vol. I. Ungar Publ. Co., New York, 1964.

[199]   Naylor, A. W., and Sell, G. R., "Linear Operator Theory in Engineering and Science."
        Holt, New York, 1971.

[200]   Neveu, J., "Mathematical Foundations of the Calculus of Probability." Holden-Day,
        San Francisco, California, 1965.

[201]   Nicolis, G., and Lefever, R., Membranes, dissipative structures, and evolution, *Advan.
        Chem. Phys.* **29** (1975).

[202]   Nitecki, Z., "Differentiable Dynamics, An Introduction to the Orbit Structure of
        Diffeomorphisms." M.I.T. Press, Cambridge, Massachusetts, 1971.

[203] Oleinik, O. A., Discontinuous solutions of non-linear differential equations, *Usp. Mat. Nauk.* **12**; *AMS Transl. Ser. 2* **26** (1957), 95–172.

[204] Oleinik, O. A., The Cauchy problem for nonlinear equations in a class of discontinuous functions, *Amer. Math. Soc. Transl. Ser. 2* **42** (1964), 7–12.

[205] Pallu De La Barrière, R., "Optimal Control Theory." Saunders, Philadelphia, Pennsylvania, 1967.

[206] Pars, L. A., "A Treatise on Analytical Dynamics." Heinemann Press, London, 1968.

[207] Pauri, M., and Prosperi, G. M., Canonical realizations of lie symmetry groups, *J. Math. Phys.* **7** (1966), 366.

[208] Pervin, W. J., "Foundations of General Topology." Academic Press, New York, 1964.

[209] Petrashen, M. I., and Trifonov, E. D., "Applications of Group Theory in Quantum Mechanics" (J. L. Martin, translator). M.I.T. Press, Cambridge, Massachusetts, 1969.

[210] Petrovskii, I. G., "Partial Differential Equations." Saunders, Philadelphia, Pennsylvania, 1967.

[211] Petryshyn, W. V., and Williamson, T. E., A necessary and sufficient condition for the convergence of a sequence of iterates for quasi-nonexpansive mappings, *Bull. Amer. Math. Soc.* **78** (1972), 1027–1031.

[212] Pierre, D. A., and Lowe, M. J., "Mathematical Programming via Augmented Lagrangians." Addison-Wesley, Reading, Massachusetts, 1975.

[213] Pipes, L. A., and Harvill, L. R., "Applied Mathematics for Engineers and Physicists," 3rd ed. McGraw-Hill, New York, 1970.

[214] Pizer, S. M., "Numerical Computing and Mathematical Analysis." Science Research Assoc., Chicago, Illinois, 1975.

[215] Poincaré, H., On a new form of the equations of mechanics, *C. R. Acad. Sci. Paris* **132** (1901), 369–371.

[216] Polak, E., "Computational Methods in Optimization." Academic Press, New York, 1971.

[217] Pollard, H., "Mathematical Introduction to Celestial Mechanics." Prentice-Hall, Englewood Cliffs, New Jersey, 1966.

[218] Pollard, H., "Applied Mathematics: An Introduction." Addison-Wesley, Reading, Massachusetts, 1972.

[219] Pontryagin, L. S., Boltyanskii, R. V., Gamkrelidze, R. V., and Mishchenko, E. F., "The Mathematical Theory of Optimal Processes" (K. N. Trirogoff, translator, L. W. Neustadt, ed.), Wiley (Interscience), New York, 1962.

[220] Porter, W. A., "Modern Foundations of Systems Engineering." Macmillan, New York, 1966.

[221] Potapova, A. F., Acceleration of the convergence of the optimum gradient method, *USSR Comput. Math. Math. Phys.* **11** (1971), 258–262.

[222] Powers, D. L., "Boundary Value Problems." Academic Press, New York, 1972.

[223] Rakitskii, J. V., Methods for successive step increase in the numerical integration of systems of ordinary differential equations, *Sov. Math. Dokl.* **13** (1972), 1624–1627.

[224] Ralston, A., and Wilf, H. S. (eds.), "Mathematical Methods for Digital Computers." Wiley, New York, 1960.

[225] Rishel, R. W., Weak solutions of a partial differential equation of dynamic programming, *SIAM J. Contr.* **9** (1971), 519–528.

[226] Rockafellar, R. T., "Convex Analysis." Princeton Univ. Press, Princeton, New Jersey, 1970.

[227] Rozdestvenskii, B. L., The Cauchy problem for quasilinear equations, *Amer. Math. Soc. Transl. Ser. 2* **42** (1964), 25–30.

[228] Rudin, W., "Principles of Mathematical Analysis," 2nd ed. McGraw-Hill, New York, 1964.

[229] Rudin, W., "Real and Complex Analysis." McGraw-Hill, New York, 1966.
[230] Rund, H., "The Hamilton–Jacobi Theory in the Calculus of Variations." Van Nostrand–Reinhold, Princeton, New Jersey, 1966.
[231] Saaty, T. L., and Bram, J., "Nonlinear Mathematics." McGraw-Hill, New York, 1964.
[232] Sagan, H., "Boundary and Eigenvalue Problems in Mathematical Physics." Wiley, New York, 1961.
[233] Saks, S., "Theory of the Integral" (Transl. by L. C. Young), 2nd ed. Dover, New York, 1964.
[234] Saletan, E. J., and Cromer, A. H., "Theoretical Mechanics." Wiley, New York, 1971.
[235] Sarachik, P. E., A computational technique for calculating the optimal control signal for a specific class of problems, *Asilomar Conf. Circuits Syst., 2nd* 1968.
[236] Scheid, F., "Schaum's Outline of Theory and Problems of Numerical Analysis." McGraw-Hill, New York, 1968.
[237] Serrin, J., On the differentiability of functions of several variables, *Arch. Rational. Mech. Anal.* **7** (1961), 359–372.
[238] Shampine, L. F., and Gordon, M. K., "Computer Solution of Ordinary Differential Equations." Freeman, San Francisco, California, 1975.
[239] Shampine, L. F., Watts, H. A., and Davenport, S. M., Solving nonstiff ordinary differential equations—The state of the art, *SIAM Rev.* **18** (1976), 376–411.
[240] Siegel, C. L., "Vorlesungen uber Himmelsmechanik." Springer-Verlag, Berlin and New York, 1956.
[241] Simmons, G. F., "Introduction to Topology and Modern Analysis." McGraw-Hill, New York, 1963.
[242] Skorokhod, A. V., "Studies in the Theory of Random Processes" (Scripta Technica Inc., translator). Addison-Wesley, Reading, Massachusetts, 1965.
[243] Smith, G. D., "Numerical Solution of Partial Differential Equations." Oxford Univ. Press, London and New York, 1965.
[244] Sokolnikoff, I. S., "Tensor Analysis, Theory and Applications to Geometry and Mechanics of Continua." Wiley, New York, 1964.
[245] Spiegel, M. R., "Applied Differential Equations." Prentice-Hall, Englewood Cliffs, New Jersey, 1958.
[246] Stakgold, I., "Boundary Value Problems of Mathematical Physics," Vol. I. Macmillan, New York, 1967.
[247] Stein, E. M., "Singular Integrals and Differentiability Properties of Functions." Princeton Univ. Press, Princeton, New Jersey, 1970.
[248] Stoer, J., and Witzgall, C., "Convexity and Optimization in Finite Dimensions I." Springer-Verlag, Berlin and New York, 1970.
[249] Street, R. L., "The Analysis and Solution of Partial Differential Equations." Brooks/Cole Publ. Co., Monterey, California, 1973.
[250] Strook, D. W., and Varadhan, S. S. R., Diffusion Processes with Continuous Coefficients, Courant Inst. Notes, New York, undated.
[251] Suliciu, I., Lee, S. Y., and Ames, W. F., Nonlinear traveling waves for a class of rate-type materials, *J. Math. Anal. Appl.* **42** (1973), 313–322.
[252] Taylor, A. E., "Introduction to Functional Analysis." Wiley, New York, 1958.
[253] Thomson, W. T., "Introduction to Space Dynamics." Wiley, New York, 1963.
[254] Toyoda, T., Necessity of complex Hilbert space for quantum mechanics, *Progr. Theoret. Phys.* **49** (1973), 707–713.
[255] Treves, F., "Locally Convex Spaces and Linear Partial Differential Equations." Springer-Verlag, Berlin and New York, 1967.
[256] Treves, F., "Basic Linear Partial Differential Equations." Academic Press, New York, 1975.

[257] Treves, F., Applications of distributions to PDE theory, *Amer. Math. Monthly* **77** (1970), 241–248.

[258] Tychonov, A. N., and Samarski, A. A., "Partial Differential Equations of Mathematical Physics," Vol. I. Holden-Day, San Francisco, California, 1964.

[259] Varadhan, S. R. S., Asymptotic probabilities and differential equations, *Comm. Pure Appl. Math.* **19** (1966), 261–286.

[260] Velicenko, V. V., On a generalization of Weierstrass' method to nonclassical variational problems, *Sov. Math. Dokl.* **13** (1972), 1593–1598.

[261] Vladimirov, V. S., "Equations of Mathematical Physics." Dekker, New York, 1971.

[262] Warga, J., Relaxed variational problems, *J. Math. Anal. Appl.* **4** (1962), 111–128.

[263] Washizu, K., "Variational Methods in Elasticity and Plasticity." Pergamon, Oxford, 1968.

[264] Weinberger, H. F., "A First Course in Partial Differential Equations." Xerox, Lexington, Massachusetts, 1965.

[265] Williamson, T. E., Geometric estimation of fixed points of Lipschitzian mappings, *Bollettino U.M.I.* (4) **11** (1975), 536–543.

[266] Williamson, T. E., Some theorems concerning the projection-iteration method, *J. Math. Anal. Appl.* **53** (1976), 225–236.

[267] Winter, D. J., "Abstract Lie Algebras." M.I.T. Press, Cambridge, Massachusetts, 1972.

[268] Yamamuro, S., "Differential Calculus in Topological Linear Spaces." Springer-Verlag, Berlin and New York, 1974.

[269] Yosida, K., "Lectures on Differential and Integral Equations." Wiley (Interscience), New York, 1960.

[270] Yosida, K., "Functional Analysis." Springer-Verlag, Berlin and New York, 1968.

[271] Young, E. C., "Partial Differential Equations, An Introduction." Allyn and Bacon, Boston, Massachusetts, 1972.

[272] Young, L. C., "Lectures on the Calculus of Variations and Optimal Control Theory." Saunders, Philadelphia, Pennsylvania, 1969.

[273] Zachmanoglou, E. C., and Thoe, D. W., "Introduction to Partial Differential Equations with Applications." Williams and Wilkins, Baltimore, Maryland, 1976.

Index